T0234731

SpringerBriefs in Applied Sciences
and Technology

Computational Intelligence

Series Editor

Janusz Kacprzyk, Systems Research Institute, Polish Academy of Sciences,
Warsaw, Poland

SpringerBriefs in Computational Intelligence are a series of slim high-quality publications encompassing the entire spectrum of Computational Intelligence. Featuring compact volumes of 50 to 125 pages (approximately 20,000-45,000 words), Briefs are shorter than a conventional book but longer than a journal article. Thus Briefs serve as timely, concise tools for students, researchers, and professionals.

More information about this subseries at http://www.springer.com/series/10618

Anton Panda · Volodymyr Nahornyi

Forecasting Catastrophic Events in Technology, Nature and Medicine

 Springer

Anton Panda ⓘ
Faculty of Manufacturing Technologies
Technical University of Košice
Prešov, Slovakia

Volodymyr Nahornyi ⓘ
Faculty of Electronics and Information
Technologies
Sumy State University
Sumy, Ukraine

ISSN 2191-530X   ISSN 2191-5318 (electronic)
SpringerBriefs in Applied Sciences and Technology
ISSN 2625-3704   ISSN 2625-3712 (electronic)
SpringerBriefs in Computational Intelligence
ISBN 978-3-030-65327-9   ISBN 978-3-030-65328-6 (eBook)
https://doi.org/10.1007/978-3-030-65328-6

This Springer imprint is published by the registered company Springer Nature Switzerland AG
The registered company address is: Gewerbestrasse 11, 6330 Cham, Switzerland

Preface

The purpose of this monograph is to present the new methodology for forecasting events and phenomena occurring in technogenic and natural environment of man as well as in the course of his illness.

The methodology is based on forecasting the individual state of the control object, which is carried out based on the analysis of the trend behaviour of the controlled parameter (symptom of the disease).

The methodology is implemented in the form of a software product intended for a wide range of users including manufacturers of various products, their consumers as well as medical workers to monitor the condition of their patients as well as for a wide range of people who independently monitor their health using fitness bracelets.

The use of software in technology in embedded control systems allows to radically increase the reliability of products reducing the number of complaints to be practically zero and thereby increasing their consumer properties. This circumstance eliminates emergency (sudden) equipment stops often leading to technogenic disasters. In this case, repairs pass from the category of restoration to the category of supporting with a simultaneous significant reduction in their number, inter-repair time lengthening and an inevitable reduction in spare parts consumption.

The presented methodology is in demand in the operation of products and structures that are unique or low series and for which there are no statistical data on the maximum allowable values of their controlled parameters.

The methodology is extremely necessary in the operation of a variety of machines, and for the first time in history, makes it possible to determine their individual resource identifying at the same time 100% of their defects.

In particular, the presented methodology allows to determine the time of the onset of a destructive earthquake, its strength and the coordinates of the epicentre as well as to forecast the time of the descent of glaciers and landslides long before the event.

The use of the methodology in medicine provides the assessment of the severity of a disease, forecast of its aggravation as well as allows to start treatment and preventive measures on time and to plan individual patient management tactics, thus improving the quality of medical services provided to the population.

The monograph provides a comprehensive presentation of information from data collection to their assessment. The monograph is useful for university teachers and students of technical faculties who are interested in new approaches and trends in his area.

The monograph was supported by grant VEGA 1/0226/21.

Vote of Thanks

Author expresses gratitude to reviewers for the valuable monograph, substantive and formal observations raise the overall level of quality publications.

Publisher/editor: Springer International Publishing, Switzerland

Edition scientific and technical literature: Monograph

Prešov, Slovakia   Anton Panda
Sumy, Ukraine   Volodymyr Nahornyi

Contents

About the Authors

Anton Panda, Prof. Ing., Ph.D. University studies—Faculty of Mechanical Engineering, TU Košice (Ing.-1987); terminated doctoral studies—Faculty of manufacturing technologies, TU Košice (Ph.D.,2002), associate professor of study branch 4.7.51 manufacturing technologies, FMT, TU Košice (assoc. prof.-2008), professor of study branch 4.7.51 manufacturing technologies, FMT, TU Košice (prof.-2015), 29 years of experience in the engineering company supplying the products for demanding automotive, also farm and agricultural industry (constructor of special-purpose machinery and equipment, systems analyst, head the department of development and technical preparation of production, methodist of statistical methods, commercial and technical director, director of quality). In the present, expertise and design activities in the field of development, production and verification of tapered rolling bearings, in the field of mounting various devices with rolling bearings for various domestic and foreign customers. Since 2008 (since 1994 external) operates as pedagogue and scientist at the faculty of manufacturing technologies, TU Košice, with the seat in Prešov, as well as an expert coordinator (auditor) of quality management systems. He is the author (co-author) of 15 monographs (nine foreign, six domestic)—of it two monographs in Springer publishing, two university textbooks (one foreign, one domestic), 16 university lecture notes, author's certificates (14), patents and discoveries (14), catalogues of bearings (2), several domestic and foreign original scientific papers in the scientific and professional journals, in Current Content Connect journals in Web of Science (14), in impacted journals and publications led in the world renowned databases (Thomson Scientific Master Journal List—Web of Science (Web of Science—47, Scopus—109) and in proceedings from domestic and foreign scientific conferences from the following areas: automobile production, manufacturing technologies, experimental methods in the manufacturing technologies, machining, development, manufacturing and verification of new products in accordance with the standards EN ISO 9001 and in accordance to the specific requirement of automobile manufacturers IATF 16 949 (ISO/TS 16 949), quality control, statistical methods and techniques of quality for the production of part, capability of machine, capability of manufacturing processes, capability of gauges and measuring equipment, technical preparation of production, product audit, system audit of quality management system, analysis of potential errors and their effect on construction (FMEA-K) and on manufacturing process/technology (FMEA-V), statistical regulation of manufacturing processes SPC, process of approval of part to the production PPAP, modern quality planning of product APQP, control plans and regulation, requirements the association of automobile manufacturers in Germany VDA 6.1, quality system requirements for suppliers of Ford, Chrysler, GM, specific requirements the using of EN ISO 9001:2015 in organizations ensuring the mass production in automotive industry ISO/TS 16949, method of Poka-Yoke, quality assurance before the mass production for suppliers of automobile manufacturers in Germany VDA 3.3, quality assurance of supplies for suppliers of automobile manufacturers in Germany VDA 2, productability, method of Global 8D (eight-step method for solving of problems), etc. At these works are registered the various domestic and foreign quotations and testimonials in the worldwide databases and solver of several projects and grant projects for engineering companies at home and abroad, solver of research tasks, author of the directives, methodological guidelines, technical regulations and other technical documentation for domestic and foreign manufacturing companies. He is the auditor of quality system management on Technical University in Košice. Active collaboration with the university workplaces is at home and abroad. He is recognized as an expert for the production of bearings in companies in Germany, Italy, China, Slovakia and the Czech Republic. As the coordinator of research collective and co-author of documentation, EFQM has won the award for improvement of performance in the competition national award of Slovak Republic for quality in the year 2010 for the Technical University of Košice. In the same competition, he has won the same award in year 2012, when the Technical University of Košice has obtained the highest score in its category. Since 2014, he has been a member of the Polish Academy of Sciences. Since 2014, he has been a member of ASME, USA.

Volodymyr Nahornyi, Ph.D. University studies—Faculty of Electronics and Information Technologies, Department of Computer Science, Section of Information Technology of Design, Sumy State University, Sumy, Ukraine (2006–2011); Master's degree in Information Technology of Design (2011); Dynamics and strength (Ph.D., 2015) and since 2016, senior lecturer in the Department of Computer Science, Section of Information Technology of Design in Sumy State University. He is the developer (co-developer) in courses such as CAD/CAM systems integration, mobile programming, methods and tools for processing visual information and technologies for creating software products. He is the author (co-author) of two monographs, author's certificates (3), patents and discoveries (8), 51 domestic and foreign original scientific papers in the scientific and professional journals, publications included in the world renowned databases (Thomson Scientific Master Journal List—Web of Science (Web of Science—six, Scopus—five)) and in proceedings from domestic and foreign scientific conferences from the following areas (22): information technologies in production, the determination of the individual resource of technical systems, the control of the dynamic behaviour of metal-processing technological systems and the forecasting of their working time and the determination of the technical state of the systems.

Symbols and Abbreviations

E_i	Value of the information signal registered at the i-th moment of the control object monitoring
E_0	Initial value of the information signal recorded during the first monitoring of the control object
$\bar{E}_{CR}^{ext}$	Value of information parameters recorded at the time of the accident (reaching the critical state of the control object)
T_{CR}	The experimental duration value of machine operation before destruction
$\psi, \alpha, \beta, \gamma, \eta$	Coefficients determined in the process of identifying a forecast model
τ	Forecast model argument
$\bar{E}^f$	Factual recorded values of the information signal E reduced to a dimensionless form
$\bar{E}^c$	Computed values of the information signal E reduced to a dimensionless form
m	Number of measurements made during the entire monitoring of the control object condition
$\bar{T}_{FOR}$	Mean time to failure forecast
σ_T	Standard deviation of the forecast
$v = \frac{\sigma_T}{\bar{T}_{FOR}}$	Forecast variation coefficient
P_{CON}	Confidence probability of the forecast
T_{FOR}^{UL}	Upper bound of the forecast with confidence probability P_{CON}
T_{FOR}^{LL}	Lower bound of the forecast with confidence probability P_{CON}
$T_{FOR}(\tau_0)$	First value of forecasting statistics
σ_{MTBF}	Standard deviation of the mathematical expectation of the forecast $\bar{T}_{FOR}$
n	Number of forecast values of operating time to failure of the control object
$\hat{T}_{FOR}(\tau)$	Current value of the mathematical expectation of the forecast
$\bar{T}_{EXP}(\tau_i)$	Moving average

$T_{MED}(\tau)$ Mathematical expectation of the current operating time τ_i and the moving average $\bar{T}_{EXP}(\tau_i)$

$P(\tau_i)$ Current value of the control object reliability
$$\left(P(\tau_i) = exp\left(-\frac{\tau_i}{\bar{T}_{EXP}(\tau_i)}\right)\right)$$

$Q(\tau_i)$ Current value of probability of reaching the limit state by the control object

$(m - n)$ Number of dots per extrapolation area

A Vibration level[mm/c]

$\bar{\gamma}$ Significance limit of the forecast model

a Model coefficient value

$\gamma_{(n-3;P)}^{table}$ Tabular value of the student's distribution selected by the number of experimental data included in the analysed sample (n is the number of experimental data, P = 0.95 - confidence probability)

$\bar{r}$ Correlation coefficient value

$r_{(n-3;P)}$ Critical tabular value of the correlation coefficient

F The Fisher's criterion

D_{COM} Total variance

D_{RES} Residual variance

E_{MED} Average value of information signal

R Determination coefficient

V_i Current value of the vibration velocity recorded on the pump bearings support

V_0 Initial value of vibration velocity recorded on pump bearing support obtained at the first measurement

ΔT Variance of the forecast of the control object operating time until it reaches the limit state

D Average deviation of computed and factual data

E_0 Value of the information parameter recorded during its first measurement

τ_0 Object operating time at the time of the first information parameter measurement

δ Exponential smoothing parameter

$A(\tau_0)$ Initial level of vibration occurred at the time of the equipment operation

A_{LL} Vibration limit

T_{OP} Actual inter-repair time

T_{IND} Individual operating time of the pump before the repair

a_{COND} Linguistic variable "state indicator"

$\alpha, \beta, \gamma, \lambda$ Exponents

a_{OP} Parameter that quantitatively describes the operating conditions of the control object

a_{CAP} Parameter that quantitatively describes the properties (potential) of the control object

$\acute{K}_A$	Relation of the current and initial levels of the information parameter
$\acute{K}_T$	Ratio of the current operating time and the forecasted operating time of the control object until it reaches its limit state
T_{RES}	Residual resource
T_{ACT}	Absolute value of the individual turbine resource
T_{RES}^{ACT}	Actual residual turbine life
T_{RES}^{FOR}	Turbine residual life forecast
E_S	Current sound level [Pa]
E_{S0}	Sound level recorded during the first measurement [Pa]
$\widehat{E}_S$	Root mean square of the useful sound signal [Pa]
$\widehat{\varepsilon}(\tau)$	Root means quare of sound interference [Pa]
S	Cutting speed [m/min]
f	Tool feed [mm/rev]
a	Cutting depth [mm]
$\widehat{E}_S^{SUM}$	Total value of the sound field in the cutting zone [Pa]
$\widehat{E}_S$	Root mean square of the useful signal generated by the given machine and each of its surrounding machines [Pa]
$\Delta\tau$	Time step of the sound signal sampling [μs]
f_{DIS}	Sampling frequency[Hz]
n_{PCS}	Number of readings of the sound signal [piece]
τ_{Sh}	Pulse duration [μs]
L	Length of the rod [m]
C	Velocity of propagation of longitudinal vibrations in the metal [m/s] ($C = 5170$ m/s)
h_r	Cutting tool wear [mm]
Δ	Radius change of the machined workpiece surface [mm]
VB	Value of chamfer wear of the major back surface [mm]
R	Correlation coefficient value
Ra	Roughness parameter [μM]
$Ra(\tau_0)$	Altitude roughness parameter defined at the start of the cutting process [μM]
$Ra(\tau)$	Altitude roughness parameter defined at the current time of the cutting process [μM]
[VB]	Maximum allowable wear rate on the rear surface of the cutting blade tool [mm]
γ	Tool wear rate
T_{IND}	Individual tool life [min]
T_{REQ}	Duration of the part (the duration of the passage) required by the technical process [min]
D_D	Workpiece diameter [mm]
L_D	Length of the machined surface [mm]
S_0	Initial (set by technological process) value of cutting speed [m / min]

f_0	Initial (specified by technological process) feed rate [mm / rev]
E_S^C	Estimated sound trend value [Pa]
E_{S0}^C	Initial estimated sound trend value [Pa]
τ	Current operating time of the cutting tool [min]
x,y	Exponents
E_S^f	Actual value of the trend (time series) of the sound generated during edge cutting machining [Pa]
k	Number of measurements taken during the entire monitoring of the cutting tool
P_{RP}	Pressing force acting in a friction pair [N]
V_{SLIP}	Relative sliding speed of friction pairs [m / min]
σ	Mechanical stress[Pa]
σ_{-1}	Fatigue limit [Pa]
N_6	Base cycles
N	Operating time in cutting tool cycles before failure
M	Fatigue curve exponent
$P_{x,y.z}$	Component cutting forces [N]
C_p	Constant for the given type of machining, machined and tool materials
K_p	Correction factor
$S(\tau)$	Adjustable (optimal) cutting speed [m / min]
$f(\tau)$	Adjustable (optimal) feed rate [mm / r]
r_B	Cutter blade tip radius [mm]
T_{PAS}	Machine time required to complete the current passage [min]
HR	Frequency component that corresponded to the heart rate [Hz]
A_{HR}^C	Calculated values of the HR amplitude [μV]
A_{HR}^{ACT}	Actual values of the HR amplitude [μV]
L	Epicentral distance
T_{REC}	Signal recording time
f_{ROT}	Earth's rotation frequency
f_{AM}	Modulating signal frequency
f_{MT}	Moon tides frequency
A_{MT}	Standard seismic signal tidal harmonic amplitude
A_{SW}	Tidal harmonic amplitude of a seismic wave
T_{AM}	Amplitude modulation period
T_{ROT}	Earth rotation frequency period
T_{MT}	Period frequency of semi-daily Moon tides
T_{UB}	Upper bound of the forecasted time range
T_{LB}	Lower bound of the forecasted time range
T_{MPV}	The most probable forecast time value
T_{CTF}	Calendar time forecast
M_S	Earthquake magnitude, adopted as a standard
A_{MT}^C	Calculated value of the tidal harmonic amplitude
A_{MT}^S	Standard value of the tidal harmonic amplitude

T_{ACT}	Actual time of earthquake occurrence
σ_F	Standard deviation of the forecasted value
σ_E	Standard deviation of the forecast of longitude coordinate E
σ_N	Standard deviation of the forecast of latitude coordinate N
σ_M	Standard deviation of the forecast of magnitude M
N_F	Forecast of latitude coordinate N
N_{ACT}	Actual latitude coordinate N
E_F	Forecast of longitude coordinate E
E_{ACT}	Actual longitude coordinate E
M_F	Forecast of magnitude M
M_{ACT}	Actual magnitude M
N	Number of seismic stations
A_{SW11}	Tidal harmonic amplitude of the reference earthquake of 2011
A_{SW19}	Tidal harmonic amplitude of the future earthquake of 2019
x_i	Geodesic latitude location of the i–th supporting reference seismic station [degrees]
x_1	Geodesic latitude of a point with a single epicentre distance [degrees]
y_1	Geodesic longitude of a point with a single epicentre distance [degrees]
$\Delta y_i = y_1 - y_i$	Difference between the geodetic longitude location of a point with a single epicentral distance and the i – th one reference seismic station [degrees]
m	Number of seismic signal measurements (number of trend members)
N	Number of reference seismic stations
p, q	Exponents ($p<q$)
γ	Experimental parameter ($\gamma \neq 0$).
$A_{MT\,i}$	Seismic signal tidal harmonic amplitude, recorded by the i-th reference seismic station in the period, preceding an earthquake ripening in a given area [μm]
A_{MTk}^S	Standard seismic signal tidal harmonic amplitude recorded by the k-th reference seismic station [μm]
L_i	Distance from the i-th reference seismic station to the centre of the maturing earthquake [km]
L_k^S	Epicentral distance of the k-th reference seismic station, used in the measurement of the reference seismic signal [km]
N_S	number of reference seismic stations used to monitor reference earthquake

Chapter 1
Introduction

The development of global civilization, aimed at improving the human life comfort on planet Earth, has led in many ways to the opposite effect. The human habitat, which is largely man-made, is becoming increasingly dangerous for man because of the unpredictability of technogenic and natural disasters accompanying human activity.

The statistics of such disasters are the evidence, and the number and severity of disaster consequences are steadily increasing.

In this regard, we present a methodology for forecasting technogenic and natural events and phenomena enabling to forecast or to interrupt this harmful chain of seemingly inevitable tragedies.

We want to emphasize that for the first time in the history of forecasting, the presented methodology allows us to determine the actual time of a forecasted event occurrence of the predicted event.

Particularly in industry, this forecasting methodology has been expected to be used at all stages, both in manufacturing and in operation, and both in large-scale and, most importantly, in unique and low-series product manufacturing. The latter is currently the most in demand, since such equipment, for example, the reactors of nuclear power plants and overpasses in megalopolises, has exhausted their resources.

However, the current methods of diagnosing the technical condition of technological systems, diverse in their design and purpose, are reduced to fixing the instantaneous technical condition of the control object. At the same time, the diagnostic procedure consists in comparing the set of controlled parameters with some standard set of parameters.

In other words, the current methods of diagnosis are reduced to fixing the instantaneous condition. Figuratively speaking, they enable to obtain an instantaneous "photograph" of a dynamically developing event or phenomenon. Then, the "photograph" is compared with some of its reference image characterizing the criticality degree of the observed phenomenon artificially stopped for a moment.

A. Panda and V. Nahornyi, *Forecasting Catastrophic Events in Technology, Nature and Medicine*, SpringerBriefs in Computational Intelligence, https://doi.org/10.1007/978-3-030-65328-6_1

However, the experience of forecasting convincingly indicates that the standards, usually representing average statistical data on the observed value (controlled parameter), correspond to the "photograph" only in general terms and therefore are ineffective. In the world of technology, standards for low-series or unique (single) technical systems do not exist at all. As a result, forecast errors in their resource may occur, leading to serious technogenic disasters.

The task of assessing the criticality degree of the technical condition of unique machines, for which no statistics on their critical state are available, is not solvable at all according to the current methodology.

This can be explained by the fact that each product, due to the existence of technological tolerances for its manufacture and assembly, is unique and therefore it is a standard for itself.

Persistent ignoring of this fact leads to unavoidable errors when making a diagnosis of the technical condition of a particular product sample. Moreover, this applies equally to objects both having and not having standards describing the criticality degree of their technical condition. First of all, it is connected with the average statistical nature of the norms, which do not take into account individual features of the equipment under control. Such uncertainty is accompanied by severe technogenic disasters.

The solution to the problem turned out to be unexpectedly simple and consists in replacing the standard of the level of the controlled parameter characterizing the static (stopped) state of the control object with the standard and characterizing the duration of the change of this state, i.e. dynamics.

As a result of the standard replacement, two time intervals are subject to comparison. The first interval characterizes the product's operating time at the moment of monitoring its technical condition, and the second one characterizes the forecast of the moment when the controlled parameter reaches its maximum permissible value.

The forecast is carried out as a result of monitoring the trend of the controlled parameter. The behaviour of this trend contains comprehensive information on individual behaviour of the control object, and therefore it is a reliable forerunner of the forecasted phenomenon.

It should be noted that monitoring the trend of the controlled parameter is the only way to solve the problem of forecasting the resource of unique machines. There are no reliable statistical data on the standards of the critical state, and therefore, according to the current methodology, the problem of forecasting is basically the impossible task in this case.

As a result, a fundamentally different methodology has been developed. This methodology radically differs from the methodology adopted for forecasting the onset of the critical state of technical systems. For the first time in the history of operating various systems that are diverse in their design and purpose, including unique systems, such as nuclear reactors, portal cranes and bridge overpasses, the presented methodology enables to determine their individual resource meeting the actual conditions of their application.

The methodology is applicable in the production process allowing, for example, for the first time in the history of machining, to determine the actual resource of the cutting tool directly during the cutting process as well as to change processing conditions, thereby eliminating the seemingly insoluble product defect problem.

According to the proposed methodology, a regression equation has been compiled between the controlled parameter and the time of its registration based on the results of monitoring the observed process. The equation is compiled in such a way that one of its coefficients is the desired time of the onset of the critical state of the control system.

The diagnostic algorithm that uses information on the actual resource is built in such a way that 100% of the defects are detected, thus eliminating the omission of the defect that is still considered to be a "scourge" of the diagnostic procedure.

Various examples of developed forecasting methodology applications are presented in this monograph.

Chapter 2
Analysis of the Current State of Forecasting Objects and Phenomena

The development of global civilization aimed at improving the human life comfort on planet Earth has led to the opposite effect. Human habitat, which is largely technogenic, is becoming increasingly dangerous for man because of the unpredictability of man-made disasters that accompany human activity. These factors increase the risk of serious diseases.

In this regard, we present a new methodology for forecasting technogenic and natural events and phenomena aimed at monitoring the condition of a wide range of industrial products including unique and low-series ones as well as monitoring the flow of technological processes and forecasting the onset of natural disasters and forecasting the time of human disease exacerbation.

The forecasting methodology is based on the analysis of behaviour of the trend of the controlled parameter in the observed period regardless of its origin physics.

2.1 Problem Statement

The presented methodology allows for the first time in the history of forecasting to determine the time of occurrence of a forecast event or phenomenon based on the natural course of the observed process, thus excluding average statistical data on its critical state.

Particularly in industry sector, this forecasting methodology has long been expected to be used at all stages of production and operation, both large-scale and, most importantly, unique and low-series products. For example, reactors of nuclear power plants and overpasses in megalopolises have exhausted their resource.

Equally, this applies to natural events and phenomena, the frequency of occurrence of which is constantly decreasing and the severity of the consequences increases.

A. Panda and V. Nahornyi, *Forecasting Catastrophic Events in Technology, Nature and Medicine*, SpringerBriefs in Computational Intelligence, https://doi.org/10.1007/978-3-030-65328-6_2

2.2 Current State of the Problem

The current methods of forecasting in the area of technology, seismology and medicine are reduced to fixing the instantaneous condition of the monitored object, event or phenomenon (the patient's condition). In this case, the procedure for assessing the criticality degree of the controlled object condition consists in comparing the controlled parameters (symptoms) with some standard value called the norms [1].

In the world of technology, there are no statistically reliable norms for low-series or unique (single) technical systems. In the process of responsible technical system operation, there may occur errors in resource forecast of similar objects resulting in severe technogenic disasters.

Resource forecasts and evaluation of criticality degree of product condition, for which such norms formally exist, may contain similar errors. This is explained by the fact that the existence of technological tolerances for the product manufacture and assembly makes it, to a certain extent, unique and therefore the norms relate to it only with a certain probability. At the same time, the norms have an average statistical character and do not take into account individual characteristics of the monitored equipment. This fact only aggravates the problem.

Ensuring the product quality during operation is unthinkable without knowledge of their actual resource. The availability of this information allows us to give up using the products according to the average schedule of preventive maintenance and begin to use the machine according to its actual condition.

Information on the actual resource is the basis of modern technical machine diagnostics aimed at detecting defects at the earliest possible stage of their development, long before failure appearance. The solution to this problem is not possible without the use of advanced methods for forecasting the technical condition of products.

Resource forecast allows us to:

- determine the moment of timely and actually necessary stop of the monitored object for the repair;
- purposely regulate the duration of inter-repair time excluding sudden emergency installations;
- reduce the time of repair downtime due to previously obtained information about the reason for the repair stop.

The objective of forecasting is achieved by determining the degree of operability (degree of criticality of the technical condition) of the product based on the information signal received from the product. The degree of criticality, in turn, is closely correlated with the degree of exhaustion (development) of a product resource equal to the time it takes for the product to reach its maximum permissible condition.

It should be noted that the currently adopted method of resource forecasting, which consists in extrapolating the graph of the analytical dependence (forecast model) until intersection with the maximum permissible level of the information signal generated by the product, is far from perfect. This is due to the fact that information on the limit level is either not available or applies to a rather limited range of products for which similar data could be available in the form of "Vibration Standards" [1].

However, the existence of "Vibration Standards" does not guarantee against errors in resource forecasting even for similar products for the reason that these norms were developed for machines that significantly differ from modern ones in specific load, structural materials, manufacturing technology, etc.

As a result, dynamic behaviour of modern machines with an insufficient degree of reliability is described by the existing "Vibration Standards".

For this reason, in the process of solving the problem, we had to abandon the use of any standard levels of information signal and focus on analysing the behaviour of the trend of the information signal which always accompanies the product operation. The physical nature of the signal does not play a significant role. The main requirement for the signal concerns the degree of sensitivity of the product behaviour to changes due to the changes in its technical condition over time.

The essence of the method lies in the fact that according to the results of regular tracking of information signal, its trend is compiled, which is the source material for the parametric identification of the trend model (hereinafter referred to as the forecast model). The model is composed in such a way that numerical value of the desired resource of the control object is calculated on the basis of the coefficients of the forecast model determined in the process of the indicated identification.

At the same time, the forecast model itself with a high degree of statistical significance and reliability describes the behaviour of trends of information signals of a different physical nature, including medical signals. Consequently, this circumstance provides the necessary degree of statistical significance and reliability of determining the coefficients of the model and, consequently, the desired resource. This is especially important for medical field because of individual characteristics of man's body.

Reference

1. ISO 10816–1: Mechanical vibration. Evaluation of machine vibration by measurements on non-rotating parts (1995)

Chapter 3
Specification of Problem Solutions

3.1 General Statements

The solution to the forecasting problem is to replace the standard level of the controlled parameter characterizing the averaged static condition of the control object with a standard characterizing the nature of change in this condition, i.e. its dynamics [1]. As a result of this replacement, not the actual and normalized levels of the controlled parameter are compared, but two time intervals, the first of which characterizes the current duration of observation of the event or phenomenon, and the second—the forecast of the duration of achievement by the monitored event or phenomenon of its maximum permissible critical state.

The forecast is carried out in the process of identifying the trend model of the controlled parameter. One of the coefficients of the model is the desired duration of the achievement by a monitored event or a phenomenon of its maximum permissible critical state. The development of the forecast model is discussed in this section.

3.2 Forecasting Tool

In world practice, according to ancient Greek, the science that deals with forecasting is called *prognostics* (prógnosis—forecast, prediction). *Prognostics* deal with theoretical and practical issues in the broad sense of the word. At the same time, prognostics are the science dealing with laws and ways of developing forecasts [1]. The objective of *prognostics*, as a scientific direction of engineering activity, is to forecast, i.e. to forecasted possible results of the phenomenon under consideration.

In English, the synonym for prognostics is the term *forecasting*. The main goal of *forecasting* is to develop special forecasting methodology in order to increase the effectiveness of methods and techniques for developing forecasts.

© The Author(s), under exclusive license to Springer Nature Switzerland AG 2021
A. Panda and V. Nahornyi, *Forecasting Catastrophic Events in Technology, Nature and Medicine*, SpringerBriefs in Computational Intelligence,
https://doi.org/10.1007/978-3-030-65328-6_3

The main goal of *forecasting* is to develop special forecasting methodology in order to increase the effectiveness of methods and techniques for developing forecasts.

Scientific forecast is a multivariate consideration about the possible outcomes of the existing situation with an assessment of the probability of implementation of each of the possible options.

Forecasting is based on the three complementary sources of information about the future:

– evaluation of development perspectives and the future condition of the forecasted phenomenon based on experience, mostly using analogy with the well-known similar phenomena and processes;
– conditional continuation into the future (extrapolation) of trends, development patterns of which in both past and present are quite well known;
– a model of the future condition of a phenomenon, for example, a process constructed in accordance with the expected or desired changes in number of conditions whose development perspectives are well known.
– In this regard, there exists a number of complementary ways to develop forecasts:
– *interpolation*—identifying of intermediate value between the two known moments of the process;
– *extrapolation* (forecasting itself), i.e. identification of future values in the forecast segment outside the process under consideration;
– *building* a dynamic series of information signal value development during the forecast time base in the past and in future (retrospection and forecast prospectus);
– *modelling* for building search and normative models, taking into account probable or desired change in the forecasted phenomenon for the ranges of the forecast based on available direct or indirect data as well as the scale and direction of changes.

Forecast models in the form of systems of equations are considered to be the most effective ones. However, other possible types of model, in the broad sense of this term, are of great importance [2]. The presented division of forecasting methods is relative as they overlap and complement each other [1].

The task of forecasting is to study the nature of changes in the analysed indicators, i.e. the study of their dynamics. This problem is solved by analysing a dynamic series. A dynamic series is a series of numerical values of the analysed indicator which characterizes a change of control object condition arranged in a chronological sequence. If these changes occur in time, then dynamic series is called a time series.

In each time series, there are two main elements—time and specific value of the indicator (levels of time series).

Levels of time series are indicators whose numerical values make up a time series. *Time* is the moments or periods to which the levels belong.

Measurement and analysis of a time series allow us to identify development patterns of a control process over time, which, as a rule, can clearly manifest itself not at each specific level but only in a trend when analysed in a fairly long-term dynamics. *Random processes are superimposed on the basic dynamic pattern. Identifying the main trend in changing the levels, called a trend, is one of the main tasks of time series analysis.*

Forecasting is reduced to processing of recorded signals, such as sound or vibration, accompanying the operation of a machine. These signals are called observables (diagnostic parameters), and the research method is called dynamic system reconstruction.

Generally, an observable quantity is understood as a sequence of values of a variable recorded continuously or at certain intervals of time. Quite often, the term "time series" is used instead of the term "observable".

The objective of forecast is to forecast the future values of the measured characteristics of the object under consideration based on observations. The task is solved by determining the parameters of the approximating function. This function recursively sets the value of a time series according to its several previous values. Thus, forecasting of a mechanical system behaviour, the coefficients of which, as a rule, are not known for certain, is proposed to be replaced by forecasting of the dynamics of a time series it generates.

Time series can be considered as a sample implementation of amplitudes of the information signal from their infinite population, generated during the operation of a machine [3]. Figure 3.1 presents an example of a time series accompanying the operation of the pump. The length of time, for which it is necessary to determine the values of a time series, is called time of forecast [1].

3.3 Time Series Forecasting

Speaking of time series forecasting, it is necessary to distinguish two interrelated concepts—the forecast method and the forecast model.

The forecast method is a sequence of actions that must be performed to obtain a time series forecasting model.

The forecast model is a functional representation that adequately describes a time series and is the basis for obtaining future process values. Speaking of forecast models, the term "extrapolation model" is often used. If the type of the approximating function is known, then the problem is reduced to finding the coefficients included in it.

The purpose of creating a forecast model is to obtain such a model for which the mean absolute deviation of the true value from the forecasted value tends to the minimum for the specified time of forecast. After a time series forecast model is defined, it is required to calculate the future values of the time series as well as their confidence intervals.

In the behaviour of a time series, two main trends are revealed—*the trend* and *fluctuations around the trend*. Fluctuations around the trend can be random or deterministic. In accordance with this fact, methods for analysing or forecasting a time

Fig. 3.1 Typical curve of
mechanical wear of friction
pairs: **a** real curve and
b wear curve model

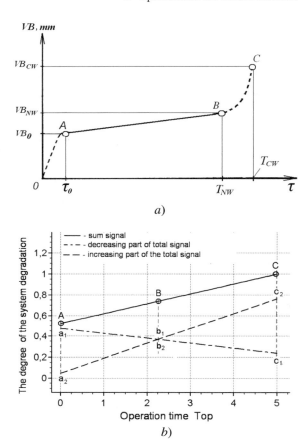

a)

b)

series are divided into random (statistical) and deterministic. In the process of statistical analysis, the statistical coefficients of the series: *mathematical expectation, the root mean square value and other coefficients of a random process* are determined.

In a deterministic analysis, the type of analytical dependence is determined and its graph describes the trend change over time most closely. To determine the parameters of the analytical dependence, as a rule, least square deviation method is used. The quality of the trend description is quantitatively characterized using a number of criteria—the Student and the Fisher criteria, etc. [4].

In practice, two methods for analysing a time series are combined, describing the trend by analytical dependence with deterministic parameters, and the forecasted value of a series is supplemented with boundary values covering it with a certain confidence probability. The boundary values are calculated from the results of statistical analysis of random deviations of a series from the trend.

Even with well-known forecast models $E = E(\tau)$, formulas describing these models can be very complex and unsuitable for practical use both in the process of mathematical analysis of physical data and in applied problems, especially in calculating the expected measurement results and in mathematical modelling of physical processes. In addition, practical registration of physical data is performed, as a rule, with a certain error or with a certain level of noise, which according to their values can be higher than the theoretical error of signal forecast when calculating using complex, although very accurate formulas. It does not make sense to design systems for processing and analysing signals using high-precision formulas if increasing the accuracy of calculations does not have the effect of increasing the accuracy of data processing. In all these conditions, the problem of approximation arises.

Approximation is a representation of complex functions $E(\tau)$ or discrete samples of these functions $E(\Delta\tau)$ by simple and convenient for practical use functions of approximation $f(\tau)$ in such a way that the deviation f (τ) from $E(\tau)$ in its range of definition was the smallest according to a certain approximation criterion.

Measurement results may contain errors (inaccuracies). In this case, it is advisable to apply approximation of the source data by least square method. The choice of the approximating function is largely determined by *physics of the described process*. If the type of the approximating function is known, then the problem is reduced, as already noted, to finding the coefficients included in it.

3.4 Forecast Model Development

The issue of monitoring mechanical system condition is a rather complicated technical task. On the one hand, the process of machine operating is very complex, difficult to be mathematically described [5], and on the other hand, a model with a minimum of parameters is needed to ensure the efficiency of diagnosing its technical condition. In such cases, it is recommended to use models represented by algebraic equations [1]. These models are called models of approximation type as they are obtained using various methods of approximation of the available experimental data on the behaviour of the output parameters of the object of diagnosing.

Approximation (hereinafter—forecast) model should describe the change of the amplitude of the information signal (usually vibration or sound) over time generated during the mechanical system operation and its structural elements reacting to wear and destruction of its structural elements.

3.4.1 Development of the Analytical Expression of a Forecast Model

A typical curve, describing mechanical wear of friction pairs, is shown in Fig. 3.1a). It is characterized by the three periods: the run-in period (section OA), the normal wear period (stationary section AB) and the catastrophic wear period (section of increase BC).

According to the literature as well as practical experience, the trend of the information signal accompanying operational process of a mechanical system follows the nature of this curve.

In this connection, the task was to develop a forecast model that could adequately describe the variously shaped information signal curves.

Let us describe analytically the behaviour of the trend of the total signal ABC (Fig. 3.1a) and its components a_1c_1 and a_2c_2 (Fig. 3.1b). When developing the model, it was assumed that the information signal curve in Fig. 3.1 was the sum of two signals. These are simplified in Fig. 3.1b in the form of straight lines [6, 7]. These addends react to the various processes of degradation of the controlled system, for example, the wear of rubbing pairs and the development of cracks in power elements.

The conclusion of the analytical expression for the forecast model is based on the postulate of physics, formulated by C. Shannon: "the basic patterns observed in the past will be preserved in the future" [8]. By the way, extrapolation forecasting methods are based on this statement.

The trend of the information signal, geometrically similar triangles, is highlighted in grey (Fig. 3.2). Triangles No. 1 correspond to the approximation area (solid line), and triangles No. 2 correspond to the extrapolation section (dotted line). Let us transform the proportions shown in Fig. 3.2, obtained on the basis of the similarity of these triangles, into Eq. (3.1) for the component a_1c_1 of the trend (Fig. 3.2a) and (3.2) for the component a_2c_2 of the trend (Fig. 3.2b). At the same time, for the generality of the obtained expressions and abstraction from the specific physical nature of the

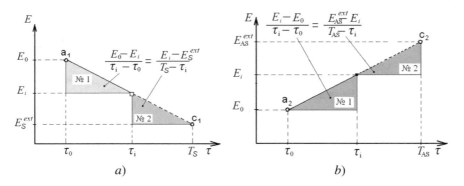

Fig. 3.2 Trend of information signal E: **a** component a_1c_1 (initial area) and **b** component a_2c_2 (ascending area)

information signal in formulas (3.1) and (3.2), a transition to the dimensionless form of the parameter $\overline{E}$ was made. The dimensionlessness is provided by normalizing the current value of the information signal E_i by its initial value E_0 recorded during the first measurement $\left(\overline{E} = \frac{E_i}{E_0}\right)$.

Let us analytically describe a similar behaviour of the trend of the information signal. Let us consider separately (Fig. 3.2) shown in Fig. 3.1 stationary (initial) area of wear AB and the area of the pre-emergency increase BC of the trend of the information signal. The run-in area OA (Fig. 3.1), due to its insignificant wear period compared to the initial section AB, is not considered. The curve of changes in the ABC information signal (Fig. 3.1) will be considered as the sum of the initial and pre-emergency sections, presenting them for simplification in the form of straight line segments starting at point A. Analytical expression for the forecast model is based on the postulate of physics formulated by C. Shannon: "the main patterns observed in the past will be preserved in the future" [9]. By the way, extrapolation forecast methods are based on this statement. In both simulated sections of the trend of information signal trend, geometrically similar triangles are highlighted in grey. Triangles №1 correspond to the approximation area (solid line), and triangles №2 correspond to the extrapolation section (dotted line). Let us transform the proportions shown in Fig. 3.3, obtained on the basis of the similarity of these triangles, into Eq. (3.1) for the initial area of the trend (Fig. 3.3a) and (3.2) for the area of ascending trend (Fig. 3.3b). At the same time, for the generality of the obtained expressions and abstraction from the specific physical nature of the information signal in the formulas (3.1) and (3.2), a transition to the dimensionless form of the parameter $\overline{E}$ is made. Dimensionlessness is provided by normalizing the current value of the information signal E_i by its

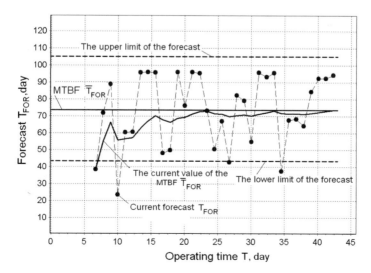

Fig. 3.3 Change in statistical parameters of the forecast during the centrifugal pump operation

initial value E_0 recorded during the first monitoring of the control system condition $\left(\overline{E} = \frac{E_i}{E_0} \right)$.

$$\overline{E}_i = 1 + \left(\overline{E}_{CR}^{ext} - \overline{E}_i \right) \cdot \left(\frac{\tau_i - \tau_0}{T_{CR} - \tau_i} \right)^{\gamma}, \quad \left(\overline{E}_{CR}^{ext} - \overline{E}_i \right) < 0. \qquad (3.1)$$

$$\overline{E}_i = 1 + \left(\overline{E}_{CR}^{ext} - \overline{E}_i \right) \cdot \left(\frac{\tau_i - \tau_0}{T_{CR} - \tau_i} \right)^{\beta}, \quad \left(\overline{E}_{CR}^{ext} - \overline{E}_i \right) > 0. \qquad (3.2)$$

The parameter $\overline{E}_{CR}^{ext}$ characterizes the amount of information parameters that are recorded at the time of the accident (reaching a critical state (crash)) of the control object, i.e. at time τ equal to the value of the T_{CR}. The exponents are also introduced into these expressions enabling to take into account the nonlinearity of the trend that is inevitably manifested in practice. From the point of view of mathematics, this fact increases the number of degrees of freedom (the number of parameters of the forecast model) and thereby increases the accuracy of graph approximation of the model of the actual trend of the information signal and, consequently, increases the accuracy of its coefficient determination when identifying the trend model. Having added (3.1) and (3.2), and entering the weighting factor ψ, which characterizes the contribution of each of the addends to their sum, we get the formula for the forecast model (3.3) of the trend of information signal:

$$\overline{E} = \left[\psi \left(1 + \eta \cdot \left(\frac{\tau - \tau_0}{T_{CR} - \tau} \right)^{\gamma} \right) + (1 - \psi) \left(1 + \alpha \cdot \left(\frac{\tau - \tau_0}{T_{CR} - \tau} \right)^{\beta} \right) \right]; \qquad (3.3)$$

Forecast model (3.3) represents the sum of two fractional rational power functions. These functions describe exacerbated modes, i.e. regimes, when the value of the controlled parameter $\overline{E}$ reaches infinity, which means the destruction of the system or a radical change in the law of its development [10]. For this reason, these fractions were used in the development of the considered forecast model, since they allow to describe dramatically varying experimental data. To implement their properties, the coefficient $\tau = T_{CR}$ was entered into their denominator. If the argument τ is equal to this coefficient ($\tau = T_{CR}$), the fractions in the expression (3.3) undergo a discontinuity. This property of the forecast model ensures its sensitivity to the trend behaviour of the information signal, forecasting the moment of its sharp change (decrease or increase), which, in turn, indicates the occurrence of an unacceptable condition of the control object. The coefficients of the forecast model are determined in the process of its parametric identification according to the results of monitoring the trend of the information signal (parameter trend $\overrightarrow{E}$). Identification consists in minimizing the difference in the graph of the forecast model (3.3) from the actual trend, carried out by searching for the minimum of the functional (3.4):

$$U = \sum_{i}^{m}\left[\overline{E}^f - \overline{E}^c\right]^2. \tag{3.4}$$

It should be noted that in practice of resource forecasting, there may be a lack of repeatability of results (insufficient forecasting stability), which is observed during the controlled period. One of the causes of instability, for example, for technical objects may be heterogeneity of the material structure of the contacting rubbing pairs, load fluctuations, etc. Let us consider the possible ways to compensate this instability.

3.4.2 Instability Compensation of Forecasting Mechanical Systems Resource

Let us consider the ways to eliminate the instability of the resource forecast of mechanical systems. A centrifugal pump will be the example of such systems (Fig. 3.3).

Statistical analysis of the forecast data, presented in Fig. 3.3, allowed us to obtain the following results:

- mathematical expectation of the resource forecast $\overline{T}_{FOR} = 73$ days;
- standard deviation of the forecast $\sigma_T = 10$ days;
- variation coefficient $v = \frac{\sigma_T}{\overline{T}_{FOR}} = 0.136$;
- upper bound of the forecast with confidence probability $P_{CON} = 0.997$

$$T_{FOR}^{UL} = \overline{T}_{FOR} + 3\sigma_T = 103 \text{ days};$$

- lower bound of the forecast with confidence probability of 0.997

$$T_{FOR}^{LL} = \overline{T}_{FOR} - 3\sigma_T = 43 \text{ days}.$$

The results of statistical analysis show that the spread of the current value of the resource forecast $T_{FOR}(\tau)$ is quite significant (43–103 days). The classic method of dealing with this phenomenon is to focus on the favourable mathematical expectation of the statistics being processed. The above is the expected value of all statistical data in general $\overline{T}_{FOR}$ collected during 43 days. The pump operation process (measuring its vibration level) was daily monitored.

In practice, the volume of the statistical database (the volume of statistics) is changed (updated), starting with the first measurement and ending with the current. In accordance with the change in the volume of statistics, its mathematical expectation should be recalculated, i.e. periodically to determine the current value of the resource $\overline{T}_{FOR}(\tau_i)$, days:

$$\overline{T}_{FOR}(\tau_i) = \frac{\left[\overline{T}_{FOR}(\tau_{i-1})(i-1) + T_{FOR}(\tau_i)\right]}{i} \tag{3.5}$$

An example of such a recalculation is shown in Fig. 3.3. In accordance with the statistical analysis laws, the current value of mathematical expectation of the forecasted value of the resource at a given time varies within

$$T_{FOR}(\tau_0) \leq \overline{T}_{FOR}(\tau_i) \leq \overline{T}_{FOR} \tag{3.6}$$

Mathematical expectation, in accordance with its properties, is characterized by a smaller variation (standard deviation σ_{MTBF}) compared with the variation of the current forecast (standard deviation σ_T):

$$\sigma_{MTBF} = \frac{\sigma_T}{\sqrt{n}} \tag{3.7}$$

Significantly less variability (inconstancy) of the current value of the mathematical expectation $\hat{T}_{FOR}(\tau)$ is clearly demonstrated in Fig. 3.3. This can be shown purely mathematically by presenting the expression (3.5) in the following form:

$$\hat{T}_{FOR}(\tau_i) = \frac{(n-1)}{n}\hat{T}_{FOR}(\tau_{i-1}) + \frac{1}{n}T_{FOR}(\tau_i). \tag{3.8}$$

The coefficients facing the components in (3.8) are in themselves weighting factors. The first coefficient at increase in the ordinal number of the current measurement (increase in the number of statistical data) tends to unity, and the second—to zero. For this reason, the influence of the newly received data on the mathematical expectation of the statistical series decreases. This property of mathematical expectation solves the problem of variability of the forecasted value of the control system resource.

However, reducing the variability of mathematical expectation automatically leads to a decrease in its sensitivity to the current change in the value of the information signal (in this case, the vibration level). The mathematical expectation disadvantage is clearly shown in Fig. 3.3.

From Fig. 3.4, it follows that after the pump has worked out 70% of its inter-repair time, the resource forecast stabilized and changed from 20 to 30 days. At the same time, mathematical expectation gave an overestimated value of the resource for at least 50 days.

To overcome this problem, a "moving average" method (exponential smoothing) was used, in which the current value of the resource forecast was determined using the following formula [5]:

$$\overline{T}_{EXP}(\tau_i) = (1-\delta)\overline{T}_{EXP}(\tau_{i-1}) + \delta T_{FOR}(\tau_i). \tag{3.9}$$

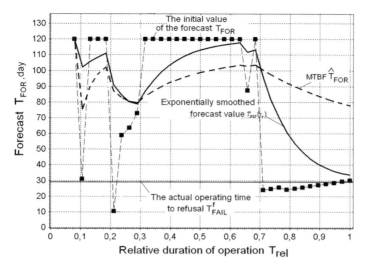

Fig. 3.4 Comparison of resource forecasts

Comparison of (3.9) and (3.8) shows that with exponential smoothing, the weighting factors of the addends are constant and do not depend on the serial number of the measurement. For this reason, the statistical series equally reacts to the current measurements and does not depend on the amount of statistical data, i.e. on the duration of monitoring a technical condition of a supervised object.

The result of exponential smoothing is shown in Fig. 3.3. As you can see, the moving average has relatively insignificant variation (variability) and, practically keeping the properties of mathematical expectation, more accurately describes the current value of the forecast resource $T_{FOR}(\tau)$ at the most responsible final section of the pump operation.

As mentioned above, the product resource is affected by many factors. As a result, the resource forecast can vary during the control object operation; therefore, the forecast should be given an interval estimate. The moving average $\overline{T}_{EXP}(\tau_i)$ is considered as the upper limit of the interval, and the most probable value of the resource $T_{MED}(\tau)$, which is found as mathematical expectation of current running time τ_i and the moving average $\overline{T}_{EXP}(\tau_i)$:

$$T_{MED}(\tau_i) = \tau_i Q(\tau_i) + \overline{T}_{EXP}(\tau_i) P(\tau_i), \tag{3.10}$$

The forecast interval between the specified boundaries represents the *forecasting field*.

The interval resource evaluation is illustrated in Fig. 3.5. For the purpose of comparison, the experimentally determined value of the actual pump resource T_{FOR}^{f} and the initial value of the resource forecast evaluation and its mathematical expectation are presented.

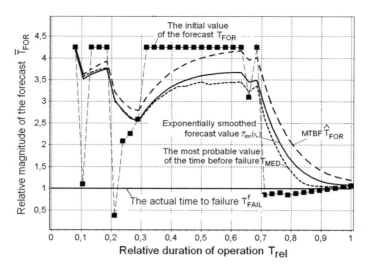

Fig. 3.5 Interval resource evaluation

All forecasted values are normalized by the actual value of the resource T_{FOR}^{f}. From Fig. 3.5, it follows that at the final stage of the pump operation, the forecast evaluation of its resource practically coincides in value with the actual resource T_{FOR}^{f}:

$$\mathrm{T}_{MED}(\tau) \approx \mathrm{T}_{FOR}^{f} \approx \mathrm{T}_{EXP}(\tau). \qquad (3.11)$$

Let us evaluate statistical significance (reliability) of the discussed evaluation of the resource forecast.

3.4.3 Evaluation of Reliability of the Mechanical System Resource Forecast

3.4.3.1 Correlation Method

From the point of view of mathematics, the resource forecast, as discussed above, consists in graph approximating of the forecast model of the actual trend of information signal. The quality of approximation is evaluated by the correlation value of the computed and factual trends. Moreover, in forecasting, correlation between computed and factual data in the forecast segment (actually forecasting) of the trend behaviour is crucial. The acceptable level of correlation in this area, in turn, determines the quality (accuracy) of resource forecasting T_{FOR}. The current value of the correlation coefficient $R(\tau)$ is determined by the following expression:

$$R(\tau) = \frac{(m-n)\sum_{i=m-n-3}^{m-n}\overline{E}^c\overline{E}^f - \sum_{i=m-n-3}^{m-n}\overline{E}^c\sum_{i=m-n-3}^{m-n}\overline{E}^f}{\sqrt{\left[(m-n)\sum_{i=m-n-3}^{m-n}\left(\overline{E}^c\right)^2 - \left(\sum_{i=m-n-3}^{m-n}\overline{E}^c\right)^2\right]\left[(m-n)\sum_{i=m-n-3}^{m-n}\left(\overline{E}^f\right)^2 - \left(\sum_{i=m-n-3}^{m-n}\overline{E}^f\right)^2\right]}} \cdot$$

$$(3.12)$$

Figure 3.6 shows the graphs of the correlation coefficients of the actual trend of the information parameter (vibration level of the pump A *mm/s*) and the graph of the forecast model (3.3). For comparison purposes, as a time parameter of the model T_{CR}, we used:

- initial forecasted value of the running time to the limit state T_{FOR}, determined while minimizing the functional (3.4);
- exponentially smoothed value $\overline{T}_{EXP}(\tau_i)$ determined by the expression (3.9).

From Fig. 3.6, it follows that the graph of correlation coefficient of the actual and computed curves, where an exponentially smoothed forecast value $\overline{T}_{EXP}(\tau_i)$ is used as a time parameter T_{CR}, is characterized by less variability and is located in the zone of high values of correlation values. Consequently, the coefficient of the forecast model T_{CR} has the same qualities of normality and reliability of its value. Actually, a similar conclusion follows from information given in Fig. 3.5.

Correlation analysis was performed under the assumption that damage to the pump components, and consequently the change in its resource, is probabilistic. At the same time, during centrifugal machine operation in general, and centrifugal pumps in particular, their damage increases, which is ultimately the reason for their repair or replacement.

Probabilistic nature of damage accumulation as well as the resource forecast is taken into account by introducing an interval evaluation of the resource forecast, the lower limit of which is equal to the most probable instant of the critical state of the control object T_{MED}, and the upper limit is equal to the exponentially smoothed

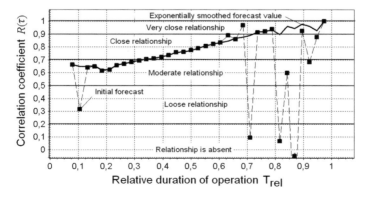

Fig. 3.6 Change in the correlation coefficients of the forecast model graph and the actual trend of the information parameter $\overline{E}$

forecast value $\overline{T}_{EXP}$. The reliability of the resource forecast is quantitatively charac-terized by the degree of compliance of the actual and computed trends of the infor-mation signal $\overline{E}$ (forecast model graph) or, in other words, the statistical significance of the forecast model.

3.4.3.2 Evaluation of Statistical Significance of a Forecast Model

Evaluation of the statistical significance of the forecast model is carried out by statis-tical evaluation of its coefficients [4]. For this purpose, it is necessary to verify the so-called null hypothesis, i.e. to check whether statistically significant evaluation of the model coefficients differs from the null. The limit of significance $\overline{\gamma}$ is established on the basis of Student's distribution:

$$\hat{\gamma} = \frac{|a|}{\sigma_a} > \gamma_{(n-3;P)}^{table},$$ (3.13)

If condition (3.13) is satisfied, then we can conclude that the considered value of the model coefficient is significantly different from the null. Evaluation of significance of the correlation coefficient between the calculated and actual data is performed according to the formula:

$$\hat{r} = \frac{r\sqrt{(n-2)}}{\sqrt{1-r^2}} \geq r_{(n-3;P)}.$$ (3.14)

If this condition is satisfied, then the hypothesis is rejected. For this, the value of the correlation coefficient is compared with its critical table value $r_{(n-3;P)}$.

To test the significance of the forecast model as a whole, the F, the Fisher's criterion, is used; for this, the total variance D_{COM} is compared with the residual variance D_{RES}. The total variance characterizes the scattering of actual data about the level of the information parameter relative to its average value:

$$D_{COM} = \frac{\sum E^2 - \frac{(\sum E)^2}{n}}{n-1}.$$ (3.15)

The residual variance characterizes the difference between the factual E^f and computed data E^c about the value of the information parameter:

$$D_{COM} = \frac{\sum (E^c - E^f)^2}{n-1}.$$ (3.16)

Fisher's criterion is found by the formula:

$$F = \frac{D_{COM}}{D_{RES}} > F_{(n-3,P)}^{TABLE}.$$ (3.17)

Fisher's criterion shows that the model forecasts the results of experiments better than average E_{MED}.

In addition, the determination coefficient R and the average deviation of the calculated data are calculated:

$$R = \sqrt{1 - \frac{D_{RES}}{D_{COM}}}, \tag{3.18}$$

$$D = \frac{\sum |E^f - E^C|}{nE}. \tag{3.19}$$

The basis for calculating the accuracy of the forecast T_{FOR} of the control object running time until its critical state is reached is an expression obtained by differentiating the expression (3.3) by the value of the information parameter $\overline{E}$ when $\tau > \tau_0$:

$$\Delta T = \frac{T_{FOR} - \tau}{\alpha E_0 \beta} \left(\frac{T_{FOR} - \tau}{\tau - \tau_0} \right)^{\beta} \sqrt{D_{RES}}. \tag{3.20}$$

The forecasted value of the resource T_{FOR} with a given probability $P = 0.95$ should be in the following confidence interval (confidence limits for the experimental sample):

$$T_{FOR} - \gamma_{(n-1;\ P)}^{table} \Delta T < T_{FOR} < T_{FOR} + \gamma_{(n-1;\ P)}^{table} \Delta T. \tag{3.21}$$

Mathematical expectation of the forecasted value of the resource $\overline{T}_{FOR}$ varies in narrower boundaries:

$$T_{FOR} - \frac{\gamma_{(n-1;\ P)}^{table}}{\sqrt{n}} \Delta T < \overline{T}_{FOR} < T_{FOR} + \frac{\gamma_{(n-1;P)}^{table}}{\sqrt{n}} \Delta T. \tag{3.22}$$

Below are the results of statistical evaluation of the quality of the forecast model of the actual trend of the information parameter by graph approximation and the accuracy of forecasting the resource of a centrifugal pump. In this case, the information parameter is a dimensionless quantity $\overline{E}$ equal to the ratio of the vibration velocity values $\left(\overline{E} = \frac{V_i}{V_0} \right)$.

The results of the statistical evaluation of the quality of the forecast model, presented in Table 3.1, indicate that the residual variance is less than the total variance; i.e. the spread of the calculated data relative to the actual is less than the dispersion of the actual data relative to each other.

Correlation and determination coefficients have high values. F, Fisher's criterion, significantly exceeds its table value. This indicates that calculated data describe the actual trend much better compared to the evaluation of the same actual data by their average value.

Table 3.1 Evaluation of statistical significance of the forecast model

The total variance D_{COM}	The residual variance D_{RES}	The determination coefficient R	Correlation coefficient $r(r_{\text{critical}})$	The average deviation D,%	Fisher's exact test, F (F_{table})
0,066	0,020	0,681	0,825 (0,370)	16,7	17,8 (2,9)

Fig. 3.7 Graph approximation of the model (3) of the actual trend of the vibration level A of the centrifugal pump rotor bearings

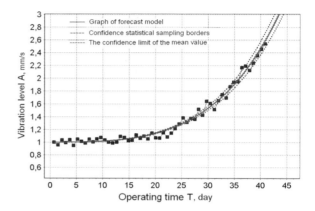

Correlation and determination coefficients have high values. F, Fisher's criterion, significantly exceeds its table value. This indicates that the calculated data describe the actual trend much better compared to the evaluation of the same actual data by their average value. This is evidenced by the insignificant absolute value deviation of confidence boundaries from their mathematical expectation and the insignificant value of the correlation coefficient r.

The high accuracy of graph approximation of the actual trend of the information parameter of the forecast model is clearly shown in Fig. 3.7.

References

1. Chetyrkin, E.M.: Statistical Methods of Forecasting, p. 200. Statistics, Moscow (1977)
2. Ivahnenko, A.G., Yurachkovskiy, Yu.P.: Complex Systems Simulation by Experimental Data, p. 120. Radio i svyaz, Moscow (1987)
3. Pronikov, A.S.: Reliability Machines, p. 592. Mechanical engineering, Moscow (1978)
4. Lvivskiy, E.N.: Statistical Methods for Constructing Empirical Formulas, p. 239. Vysshaja shkola, Moscow (1988)
5. Greshilov, A.A., Stakun, V.A., Stakun, A.A.: Mathematical Methods of Forecasting, p. 112. Radio i svyaz, Moscow (1997)
6. Lopatin, A.S.: Justification of diagnostic signs of rotor imbalance vol. 2, pp. 36–39. National Academy of Sciences of Ukraine, Institute of Electric Welding of E.A. Paton NAS of Ukraine, Kiev, Ukraine (2002). ISSN 0235-3474.

7. Barkov, A.V.: Monitoring and Diagnostics of Rotary Machines by Vibration, p. 159. St. Petersburg State Marine Technical University, St. Petersburg (2000)
8. Palmgren, A.G.: Die Lebensdauer von Kugellagern, vol. 68, No.14, pp. 339–341. Zeitschrift des Vereines Deutscher Ingenieure (ZVDI), Berlin, Germany (1924)
9. Shannon, K.: Works on Information Theory and Cybernetics, p. 830. Foreign Literature Publisher, Moscow (1963)
10. Podlazov, A.V.: Modes with an aggravation with complex indicators. Log-periodic oscillations in the fiber bundle break model, News of universities. Appl. Nonlinear Dyn. Moscow, Russia **19**(2) (2011). ISSN 15-30, 0869-6632

Chapter 4
Application of the Developed Forecasting Methodology in Various Spheres of Human Activities

4.1 Resource Forecasting in Technology

The problem of preventing failures and reducing technogenic risks is now of particular relevance. This problem is particularly acute in the operation of special-purpose technical objects and the failures of which are associated with large material losses or catastrophic consequences.

As a rule, these complex systems, which are produced in small quantities and operated in different conditions, implement extreme technologies.

The solution to the problem of preventing failures of such systems largely depends on the possibility of monitoring and forecasting their technical condition or individual resource to reduce the risk of sudden accidents and disasters.

The well-known forecasting methods [1–6] are based on experimental data approximation by a certain analytical dependence followed by approximation to the moment of intersection of its graph with the maximum permissible level of the controlled parameter according to the norms.

The decision is made by direct comparison of the controlled parameters with the limits of the operational area. The degree of their remoteness from the limits of allowable changes at the time of monitoring is used to evaluate the current value of the individual resource. If norms are unavailable, such evaluation is not feasible.

A number of examples of application of the developed methodology for forecasting the individual resource of technical systems are clearly shown in Fig. 4.1.

Let us take a closer look at the process of the individual resource forecasting for both large-scale and low-series products. In the process of forecasting, the following tasks are being solved:

- Individual resource of the technical system is determined at the moment of monitoring of its condition.
- Technical condition of the system at a given point in time is estimated.

A. Panda and V. Nahornyi, *Forecasting Catastrophic Events in Technology, Nature and Medicine*, SpringerBriefs in Computational Intelligence, https://doi.org/10.1007/978-3-030-65328-6_4

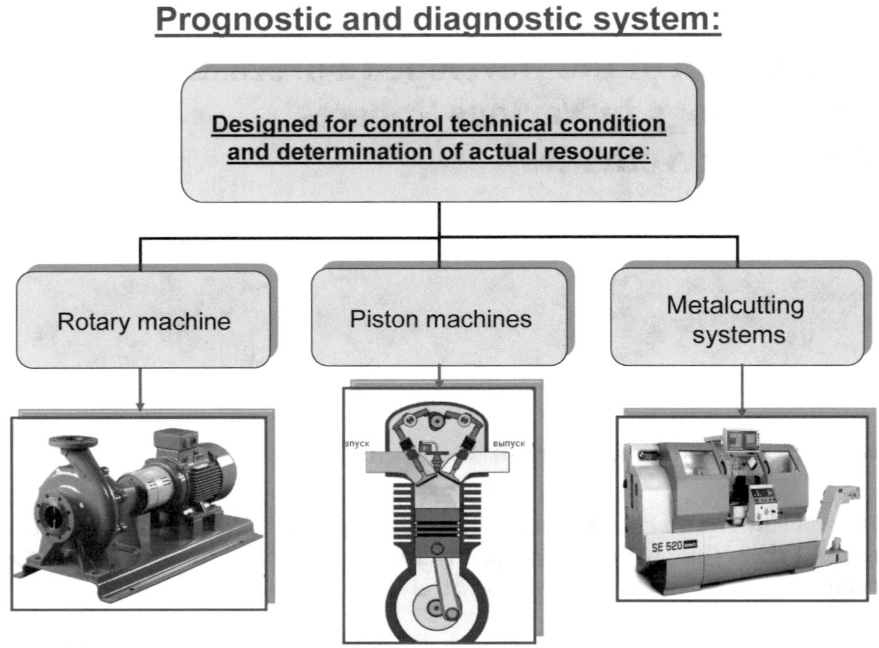

Prognostic and diagnostic system:

Designed for control technical condition and determination of actual resource:

Rotary machine

Piston machines

Metalcutting systems

Fig. 4.1 Application areas of the new forecasting methodology in technology

– Changes in technical condition are forecasted until its maximum permissible characteristics are achieved.
– Operation strategy, which ensures operability preservation of the monitored equipment during the specified period of its use, is selected.

4.1.1 State of the Forecasting Problem in Technology

Researchers have addressed the problem of forecasting the individual resource of various technical systems, including those produced in small series [1–3]. According to the results obtained, in the course of operation of similar in class-controlled objects, a collection of statistical information on the maximum allowable interval of change of the controlled parameter is performed. Then, during the equipment operation, correspondence between the forecast value of the controlled parameter and the allowable interval of its change is examined. If it enters the allowable interval, the object is considered to be operational, otherwise its future failure is recorded.

The disadvantage of this method, especially in relation to unique products or products produced in small batches, is obvious. However, it does not enable to compile sufficiently representative statistics of their failures by defining the allowable interval of the controlled parameter change.

The paper [7], which deals with monitoring of the technical condition of the ship equipment, proposes to determine the parameters of the linear equation describing the trend of the logarithm of the vibration level of the equipment. At the same time, equipment operating time, until the trend intersects with the maximum allowable vibration level of the A_{LL}, was considered as the desired resource.

If there is a lack of required amount of statistics, it was proposed to apply the "6 dB rule", according to which the vibration level A_{LL} is taken as the limit, which is 6 dB higher than the initial $A(\tau_0)$ occurring at the time when the equipment began to operate and worked properly, as it was expected. In transition to absolute values, this means that according to the "6 dB rule" between the initial $A(\tau_0)$ and the maximum allowable A_{LL} vibration levels, there exists the following relationship $A_{LL} \approx 2 \cdot A(\tau_0)$.

The disadvantage of this method is that the "6 dB rule", being purely evaluative, does not provide the necessary degree of validity and reliability when determining the individual maximum allowable load level for a specific product sample.

Further solutions to this problem are presented in the paper [8] in the example of forecasting the resource of centrifugal pumps. The results of measuring the vibration of the pump housing are presented in the form of the trend, describing the tendency of vibration amplitude change during the pump operation time.

The current trend in forecasting a machine resource, called a trend analysis [9], makes it possible to estimate the probability of deviations of the controlled parameters that differ from the values corresponding to the properly functioning monitored equipment. When trends are detected, technical condition is forecasted to evaluate the possible occurrence of faults at a given time interval. Trends are built using a computer monitoring system, with threshold values and combinations of diagnostic features previously entered into it, and resource forecast is carried out based on a comparison of current and threshold values of the controlled parameters.

The disadvantage of a trend analysis lies in the fact that for unique technical systems or small-series products, where it is necessary to apply methods for an individual resource forecasting, it is impossible to establish threshold values of the monitored parameter due to the lack of a required amount of statistics.

A common disadvantage of the existing forecasting methods [10–12] is their orientation towards statistically average normative values of the controlled parameter that do not reflect individual properties of the controlled object.

As a result, by determining the moment when the trend reaches its maximum allowable level, resource forecasting leads to errors of either the first or second kind. Both are equally unacceptable.

The solution to the problem of determining the individual resource should be sought in the way of analysing the behaviour of the trend of the controlled parameter. The behaviour of this trend for each of the control objects is unique and highly individual. Examples of similar technology are discussed below.

4.2   Forecasting the Resource of Large-Scale Products Using a Centrifugal Pump as an Example

The purpose of the research was to forecast the moment the pump is stopped for repair due to resource exhaustion of the ball-bearing-supported rotor.

The subject of investigation was the K 8/18 pumping unit (Fig. 4.2, Table 4.1), which is a centrifugal, horizontal, cantilever, single-stage pump based on the pump casing [13].

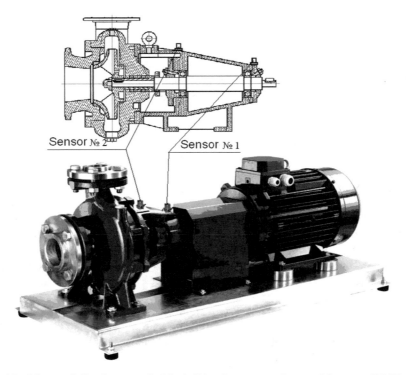

Fig. 4.2 Scheme of vibration control of the ball-bearing-supported rotor of the pump K 8/18

Table 4.1 Technical characteristics of the pump

Pump Brand	Serve, m³/hour	Head available, m	Rotation frequency, Hz	Power consumption, kW	Suction reserve, m
К8 /18	8	18.8	2900	1.20	2.80

Fig. 4.3 Graph approximation of the forecast model of the actual trend of the vibration velocity of the ball-bearing-supported rotor (sensor number 2)

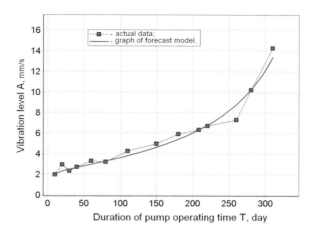

The research methodology consisted in step-by-step forecasting of the pump resource based on the results of periodic monitoring of the vibrations of the ball-bearing-supported rotor (Fig. 4.2). The measurement frequency ranged from 10 to 30 days.

Measurements were discontinued at the time of the coincidence of the resource forecast with the actual inter-repair operating time of the pump ($T_{OP} = 310$ day). Vertical vibration in the frequency ranged from 1 to 2000 Hz of the ball-bearing-supported rotor (Fig. 4.3) was measured using vibration sensors No. 1 and No. 2.

The signal from the sensors has been sent to the "sound card" of the computer, where it was digitized by the sampling frequency of 11,025 Hz. The digitized signal was subjected to further processing in order to determine the total root mean square level of vibration velocity A, mm/s and, on this basis, to forecast the pump resource. Processing was carried out according to the program compiled in the algorithmic C++ language [13] reflecting the algorithm of the forecast method discussed in Sect. 3.4.

Forecast accuracy was evaluated by the.

- degree of correlation of the forecast model graph and the actual trend of the vibration level;
- fact of getting the individual pump resource (operating time before repair)T_{IND} into the forecast field $\left(\overline{T}_{EXP} > T_{IND} > T_{MED}\right)$;
- results of the pump repair and fault detection of its ball bearings.

Measuring carried out by sensor No. 2, installed at the most loaded point of the pump (Fig. 4.2,) served as initial information for forecasting the pump resource (Table 4.2).

It follows from Fig. 4.3 that the graph of the forecast model with a sufficient degree of accuracy approximates the actual trend of the vibration level of the ball-bearing-supported rotor, as evidenced by their cross-correlation coefficient $R = 0.906$.

Table 4.2 Vibration levels recorded by sensor No. 2 (Fig. 4.2)

Hours used T, day	10	20	30	40	60	80	110	150	180	208	220	260	280	310
Vibration levels A,mm / s	1,9	2,9	2,3	2,7	3,3	3,2	3.3	3.9	4.9	6.3	6.7	7.3	10.2	13.3

Figure 4.4 clearly shows that approximately from the middle of inter-repair period (150 days), the actual resource is within the boundaries of the forecast field indicating the effectiveness of the considered forecasting method. It is very important when operating pumping equipment for it allows us to plan repair activities in advance.

Figure 4.5 shows the hearths of pitting on the inner ring of the ball bearing identified during the pump repair. As you can see, the inner ring had an unacceptable degree of pitting, which was the cause of increased vibration of the rotor support and required the pump to be stopped for repair. The bearing was located in the support

Fig. 4.4 Relationship between the individual resource T_{IND} and its forecast

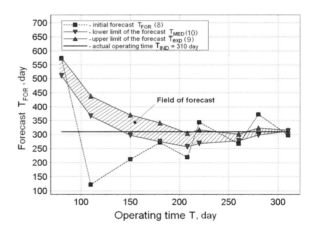

Fig. 4.5 Bearing inner ring defect

of the pump rotor, above which the vibration sensor No. 2 was installed (Fig. 4.1). This sensor has recorded the increased vibration level (Table 4.2).

4.3 Forecasting the Resource of Small-Scale Products Using a Hydro Turbine as an Example

Forecasting the resource of small-scale products was carried out in the example of the hydro turbine of the Sayano-Shushenskaya HPP. The accident at this hydroelectric power plant led to great technogenic disaster. According to the considered forecasting methodology, in the process of retrospective parametric identification of a trend model (3) of a monitored parameter (vibration level of turbine supports), an individual turbine resource was determined (operating time before the accident T_{FOR}). The identification was carried out by minimizing the deviation (4) of the calculated and actual values of the trend of the monitored parameter.

Based on the individual resource, the residual life T_{RES} was determined from the following expression:

$$T_{RES} = T_{FOR} - \tau. \tag{4.1}$$

In the process of exhaust of its resource, the product technically degrades and gradually passes a series of increasingly deteriorating technical conditions.

To formalize the description of these states, let us use the basic assumptions of the theory of "fuzzy sets" [14]. According to this theory, the criticality degree of a product technical condition can be assessed by means of the linguistic variable a_{COND}, called the "state indicator" of the test object.

This indicator, in particular, allows to compensate the lack of large volume of statistics on the maximum allowable values of the controlled parameters for small-scale products. For unique (single) products, there are no statistics at all.

The "membership function", by means of which this indicator is calculated, transforms the values of the input variables characterizing the operational load and the individual properties of the control object into the values of linguistic variables. These variables are compared with their standard values—"the terms", representing diagnoses of the current condition of the object being diagnosed [15].

4.3.1 Derivation of the Analytical Expression for the "Membership Function"

To obtain an analytical expression for the "membership function", by means of which the linguistic variable "condition indicator" a_{COND} is calculated, we will use the methods of the similarity and dimensional theory [16]. According to this theory, we assume that the indicator a_{COND} is described by the following set of defining parameters: the initial $A(\tau_0)$ and current vibration levels $A(\tau)$, the current operating time of the monitored object τ and the numerical value of its resource T_{CR}.

Assuming without loss of generality of the resulting expression that the amplitude of vibration is a vibrational shift, let us consider two dimensions—the length L ($[A(\tau_0)] = L$, $[A(\tau)] = L$) and time T ($[\tau] = T$, $[T_S] = T$). The unknown indicator a_{COND} is represented as a function of the product of these defining parameters, each of which is raised to its power:

$$a_{CQND} = f(A(\tau_0), A(\tau), \tau, T_{FOR}) = A^\alpha(\tau_0) \cdot A^\beta(\tau)\tau^\gamma \cdot T_{FOR}^\lambda. \qquad (4.2)$$

Next, we determine the values of the exponents. To do this, we replace the values in the expression with their dimensions and substitute them into formula (4.3):

$$1 = L^\alpha L^\beta T^\gamma T^\lambda = L^{\alpha+\beta} T^{\gamma+\lambda}. \qquad (4.3)$$

Since the unknown indicator is a dimensionless quantity, its dimension, according to the dimension theory, is equal to one. Accordingly, the following conditions for the exponents should be met:

$$\alpha + \beta = 0, \gamma + \lambda = 0. \qquad (4.4)$$

We supplement this system of two equations with four unknown to four equations conditions $\beta = 1$, $\gamma = 1$. The joint solution for the four equations leads to the following result: $\alpha = -1$, $\beta = 1$, $\gamma = 1$, $\lambda = -1$.

Consequently, the formula (4.6) is written in the following form:

$$a_{COND} = \frac{A(\tau) \cdot \tau}{A(\tau_0) \cdot T_{FOR}}. \qquad (4.5)$$

During its operation, technical condition of the control object changes, passing through a series of consecutively deteriorating technical conditions.

Accordingly, the actual value of the linguistic variable a_{COND}, calculated from the measurement results of the controlled parameter and resource forecast of the control object, is being changed.

The diagnosis of the technical condition should be made by comparing the actual value of the variable a_{COND} with some predetermined normative limits that separate one diagnosis from the other.

4.3.2   Determination of Normative Boundaries of the Linguistic Variable a_{COND}

Let us consider separately the numerator and denominator of the "membership function" (4.5). The numerator, which is equal to the product of the current vibration level and the current operating time (4.6), characterizes operating rate of the equipment

$$a_{OP} = A(\tau)\tau. \tag{4.6}$$

The denominator, equal to the product of the initial vibration level and the resource (10), reflects the individual properties of the control object, its capacity indicator as well as potential and its value remains practically unchanged during the control object operation.

$$a_{CAP} = A(\tau_0)T_{FOR}. \tag{4.7}$$

During the process of the control object operation, the numerator, changing from zero (at $\tau_0 = 0$), is compared with the value of the denominator ($a_{OP} = a_{CAP}$). At this moment, the linguistic variable a_{COND}, also changing from zero, reaches unity, and the object exhausts at least its inter-repair resource.

The given value of the variable a_{COND} is critical because stopping the equipment at the moment when equality of the numerator and denominator of the "membership function" (4.5) is fulfilled avoids an accident (catastrophe), which is precisely the purpose of monitoring the technical condition of supervised equipment.

To prove this statement, we will present the expression (4.5) as a product of two fractions:

$$a_{COND} = \frac{A(\tau)}{A(\tau_0)} \cdot \frac{\tau}{T_{FOR}}. \tag{4.8}$$

The first fraction is the ratio of the current and initial levels of vibration.

$$K_A = \frac{A(\tau)}{A(\tau_0)}. \tag{4.9}$$

The second is the ratio of current operating time and the resource forecast

$$K_T = \frac{\tau}{T_{FOR}}. \tag{4.10}$$

When the linguistic variable a_{COND} equals unity, the expression (4.8), taking into account the expressions (4.9) and (4.10), can be represented as follows:

$$\frac{a_{COND}}{K_A} = \frac{1}{K_A} = K_T. \tag{4.11}$$

Table 4.3 Boundary values of the linguistic variable "state indicator" a_{COND} and corresponding "terms"

«Good condition»	«Acceptable condition»	«Allowable condition»	«Invalid state»
$0 < a_{COND} < 0.41$	$0.41 < a_{COND} < 0.63$	$0.63 < a_{COND} < 1.0$	$a_{COND} > 1.0$

The parameter K_A is usually greater than unity (which corresponds to the condition $A(\tau) > A(\tau_0)$), and the parameter K_T will be smaller than unity (which corresponds to the condition $\tau < T_{FOR}$). The latter means that stopping at the moment when parameter a_{COND} reaches unity will exclude the equipment crash.

If the value of the control parameter decreases as approaching the critical state of the equipment, the value of the parameter K_T should be monitored directly, stopping the control object when the value of this parameter exceeds the threshold value of 0.9.

The range of variation of the variable a_{COND} 0…1 by analogy with the norms of vibro-activity [1] is divided into four subranges [15]. The boundaries of the subranges, also by analogy with the vibro-activity norms [1], represent the normalized series of preferred numbers $R4$. This series is a geometric progression with a root $q = \sqrt[5]{10} \approx 1.56$ [17]. The subranges characterize a gradually deteriorating diagnosis of the control object condition and are described by a standard set of verbal values —"the terms" (Table 4.3).

4.3.3   Evaluation of the Criticality Degree of the Turbine Condition

The purpose of the experiment was to present the considered methodology of resource forecasting as well as to evaluate the nature of the change of the criticality degree of the technical condition of the monitored object with time.

The subject of research was the GT-2 hydraulic turbine of the Sayano-Shushenskaya HPP (Fig. 4.6), which belongs to the category of small-scale products. Turbine suffered catastrophic destruction on 18 August 2009.

The experiment methodology consisted in retrospective forecasting of the turbine resource and evaluation of the consistent change of the criticality degree of its technical condition during the period from 21 April to 18 August 2009.

In this case, the vibration level was considered as a control parameter.

The method involved:

– building the trend of the vibration level;
– forecast of the turbine resource, carried out in the process of identifying the trend model (3), by minimizing the functional (4);
– calculation of the value of the linguistic variable a_{COND} by means of the "membership function" (9);

Fig. 4.6 Diagram of the
hydro turbine indicating the
point of its vibration
monitoring: 1—generator
bearing; 2—turbine bearing;
3—turbine impeller;
4—vibration sensor of the
turbine bearing

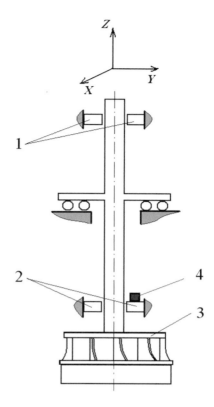

- evaluation of the criticality degree of the current condition of the turbine, carried
  out according to the results of comparison of the linguistic variable a_{COND} with
  its reference intervals ("the terms") (Table 4.3).

As the initial data for forecasting the turbine resource, the results of measurements
of the vibration of the turbine body (Table 4.4) in the area of the turbine bearing 2
were used (sensor 4, Fig. 4.6).

The measurements were carried out during the period from 21 April to 17 August
2009 [18]. In this case, the T_{CR} coefficient in the expression for the forecasting model

Table 4.4 Amplitude of vibration of the turbine bearing $A(\tau)^{*)}$, microns

Point and direction of vibration control	Date of measurement (running time τ, days, counted from 21 March 2009)					
	21 March 2009 (0)	22 June 2009 (63)	7 July 2009 (75)	23 July 2009 (91)	11 August 2009 (111)	17 August 2009 (118)
4y	90	250	500	700	800	1500

$^{*)}$ In order to increase the visibility of the demonstrated methodology, the measurement results were
interpolated to increase the amount of input data (Fig. 4.7)

Fig. 4.7 Trend of the
controlled parameter
(vibration level) and its
model

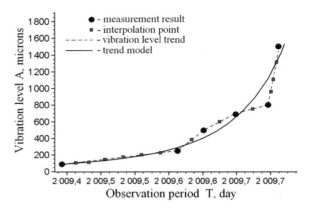

Fig. 4.8 Comparison of the
individual resource forecast
of the turbine T_{CR} and its
actual value T_{ACT}

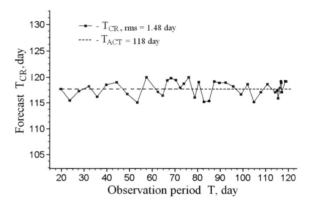

(3) was understood as the turbine operating time in days, started from 21 April 2009.
The residual resource of the T_{CR} turbine was determined by the formula (4.1).

The calculation results shown in Figs. 4.7, 4.8, 4.9 and 4.10 contain:

- the trend of the controlled parameter and its model (Fig. 4.7);
- comparison of the individual resource of the turbine T_{ACT} and its forecast T_{CR} (Fig. 4.8);
- comparison of the actual residual life of the turbine T_{RES}^{ACT} and its forecast T_{RES}^{FOR} (Fig. 4.9);
- change in the value of the linguistic variable a_{COND} until it reaches its critical value, equal to unity (Fig. 4.10).

The plots shown in Fig. 4.7 indicate a significant positive gradient of vibration
level in the precatastrophic period of turbine operation which is a reliable harbinger
of an impending catastrophe.

As follows from Fig. 4.8, the resource forecast T_{FOR}, starting 20 days from the
start of observation of the turbine vibration state, fluctuates relatively to the actual

Fig. 4.9 Change in forecasting of residual resource T_{RES}^{FOR} and its actual value T_{RES}^{ACT} during the observed period

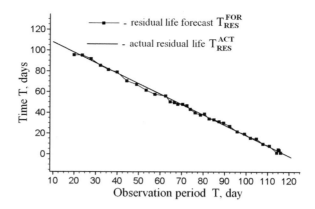

Fig. 4.10 Change in the linguistic variable "state indicator" a_{COND} in the process of monitoring the turbine condition

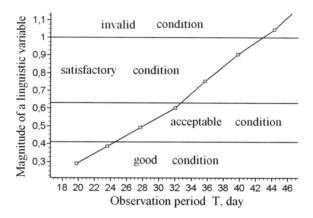

turbine operating time before the catastrophe T_{ACT} with the root mean square value rms = 1.48 days ($v = 0.0125$).

Consequently, comparison of the actual T_{RES}^{ACT} and forecasted residual resources T_{RES}^{FOR} shown in Fig. 4.9, demonstrates that they practically coincide.

From Fig. 4.10, it follows that during the process of observation, the linguistic variable a_{COND} increases in value, passing a number of its normalized values, and 44 days after the start of regular monitoring of the turbine, it reaches its critical value—the unity.

The results of the retrospective forecasting of the turbine operating time before the catastrophe clearly indicate that the use of the presented methodology for forecasting the resource of technical systems would help to avoid a catastrophe.

In reality, during the precatastrophic period, the principle of the dichotomy "fit–not fit" has been used, guided by the currently used limit values [19] ($A_{LI} = 180\,\mu m$ at the working frequency). These values do not suppose a gradation of the criticality degree of the technical condition of the turbine shown in Table 4.2.

In the case of the proposed methodology application, a stable forecast repetition of the turbine operating time until its critical condition is reached as well as the fact of change in the linguistic variable a_{COND} from the initial value to unity in a short period (44 days) would undoubtedly be the basis for stopping the turbine to ascertain the reasons of its unusual behaviour.

The striking difference between the existing and the proposed methodology for monitoring the technical systems condition, demonstrated in the example of the hydraulic turbine, indicates the long overdue need to introduce it into the practice of individual resource forecasting for various technical systems either unique or small scale.

Analysis of the current state of the problem on technical systems monitoring [20] shows that dynamics of changes in the controlled parameter and its trend have still being ignored, and monitoring itself is reduced to comparing the controlled parameter with a certain predetermined reference value, which, as we all know [15], does not exist.

4.4   Forecasting Individual Resource of the Aircraft Engine

The purpose of forecasting was to determine the operating time T_{ACT} of the engine until its resource was exhausted according to the method considered in the monograph. Experimentally obtained statistical data on vibration of the rotor of the turbocharger of the gas turbine engine TB3-117 of the Mi-8 helicopter were used as initial information for forecasting [21]. The initial data are presented in the form of the trend of the vibration level A of the engine, recorded in the frequency range of rotation of its rotor (Fig. 4.11).

It should be noted that the vibration level A of the engine, as a diagnostic sign, is highly informative and determines the condition of the engine mounts, its flow path and suspension assembly wear, imbalances, rotor misalignment, etc. [22, 23].

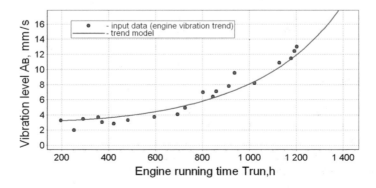

Fig. 4.11 Approximation of engine vibration data using forecast model graph

To confirm the accuracy of forecasting the actual (individual) engine resource, the magnitude of engine damage Q was determined (Fig. 4.13). The parameter Q was calculated on the basis of the Palmgren–Miner linear damage accumulation rule [24, 25] using the formula:

$$Q = \sum \frac{\tau_i}{T_{FORi}}. \tag{4.12}$$

Forecasting results are shown in Fig. 4.12 and in Table 4.5.

Forecasting (Fig. 4.12) shows that in the course of time, the engine's resource has decreased, from about 3,000 to 2,000 h at the time of the last monitoring of its

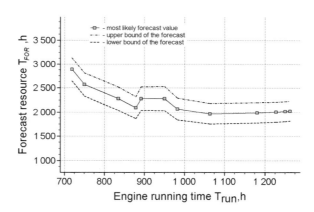

Fig. 4.12 Engine resource forecast

Table 4.5 Resource forecast T_{FOR}, time

Engine running T, time	720	834	890	1063	1260
Forecast	2641...3136	2041...2516	2038...2524	1751...2183	1812...2232

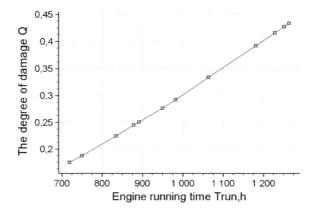

Fig. 4.13 Degree of the engine damage

condition. Accordingly, the degree of its damage Q increased, reaching the value of 0.44 at the time of its last monitoring (Fig. 4.13). Consequently, the running time, until the engine resource is exhausted, varies within 600–900 h (Table 4.5).

Similar specificity in the resource definition is extremely necessary when evaluating the criticality degree of the technical condition of such highly demanding products as aircraft engines. It allows us to make an informed decision on really necessary and timely replacement of the engine with a new one.

4.5   Forecasting of the Individual Resource of the Cutting Tool

4.5.1   General Statements

The quality of product machining by cutting must be monitored throughout the entire technological chain. The decisive influence on the quality of controlled parameters has wear of the cutting tool. This fact requires implementing an adaptive control with the varying purposes of the cutting modes that allows prolonging the defect-free performance period of a predetermined operation of machining materials by cutting.

There are two ways of controlling tool wear—direct and indirect. In industrial practice, the indirect way became widespread. In this regard, the tool wear adaptive control is indirectly evaluated by measuring information signals having various physical nature and accompanying the cutting process.

Effective performance management is dependent upon the noise resistance of the information signal and its sensitivity to the tool wear degree and the workpiece surface roughness. These properties of the controlled signal are necessary for the effective management of defect-free workpiece manufacturing because they eliminate the risk of errors.

The currently used dependencies between the information signal and tool wear [26–45] are phenomenological, which, often, does not have a quantitative justification. Therefore, the research purpose, the result of which are presented in this article, was to determine the correlation between one of the promising information signals— the sound generated during the cutting process and the tool wear and the roughness of the machined part surface.

4.5.2   Monitoring the Status of the Cutting Tool by the Sound that the Cutting Process Generates

Individual resource control (resistance) of the cutting tool is based on the measurement of information signals of different nature representing a variety of external factors: mechanical stress, vibration, the processing system elastic deformations, the electric current, chemical exposure, and the like, which have a decisive influence on the tool wear degree and, consequently, on quality of the manufactured product [26].

So, in the course of the tool wear process, such parameters as cutting force [27], the torque [28] and cutting power [29] are changing. These parameters are measured by dynamometers. Applying the vibration sensor, for example, when turning on the tool holder, the tool vibrations accompanying inevitably the cutting process are measured [30].

During the interaction of the wearing tool with a workpiece, acoustic emission waves are generated, that are recorded by acoustic emission sensors in the frequency band from 1 kHz to 1 MHz. Acoustic emission is more sensitive than the force factors and vibration to tool wear [31], but at the same time, it makes acoustic emission more sensitive to noise disturbance caused by the environment influence and the work of the machine structural units.

For this reason, the "integrated settings" registration is referred to as they are more resistant to noise disturbance. For example, thermocouples monitor the temperature in the cutting zone [32], they measure the thermo-EDS [33] and the electrical conductivity [34] of the tool–workpiece contact pair. The disadvantage of the "integral" methods is their considerable inertia and the need to incorporate a thermocouple and electrical connections into the instrument. Adaptive control also uses acoustic emission signals [25] and measurement of vibrations [36–40].

The vibrations and acoustic data signals turn out to be the simplest in the registration and subsequent processing [30, 36]. However, in practice, the use of these signals is related to a significant problem that is difficult to solve. These information signals are recorded by the contact method. The method requires implementation of a mechanical communication of the sensor with the information signal source (the object surface). To meet this requirement, it is necessary to solve two difficult tasks: to choose an informative point on the controlled objects where you want to install the sensor and to avoid interference, always related to the method of measurement.

When monitoring the cutting process, an informative point is considered to be the tool contact point with the workpiece. At this point, the useful signal carrying information about the course of the cutting process is generated. Placing the sensor at this point is impossible. Selection of other possible closest control points, for example, when the turned workpiece is on the cutting tool holder, makes it necessary to exclude the biggest interference from the measurement results. Noise disturbance, in this case, is the vibration generated by the machine operating nodes. To select the useful signal without significant distortion on the intensive level background and with the complex in frequency parasitic vibrations is almost impossible. The solution to this problem is achieved by eliminating the sensor contact with the machine tool vibrating surface that is implemented with non-contact measurement method. In this case, it is of interest to control the sound accompanying the cutting process [41–44].

Will consider quantitative degree evaluation of the relation between sound and wear.

On the surface of the part, protrusions and depressions (see Fig. 4.14), called roughness, [41] are formed because of irregularities of machine tool processing.

Fig. 4.14 Surface of the part
after its processing by cutting
magnified by a microscope

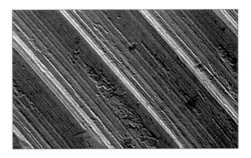

The cause of the roughness is the trace left by a tool on the machined surface caused by mutual oscillation of the tool and the workpiece. This phenomenon that is the tool interaction with the workpiece became 135 years ago the basis of recording, during which the cutter leaves a trail (a phonogram) on the lateral surface of the wax cylinder and then on the surface of flat records [42].

Figure 4.15 shows a 1000 times magnified part of the soundtrack recording, which is a part of the recorded track on a gramophone record. This soundtrack is just the surface of the part, which is formed while it is processed by a metalworking tool.

Roughness, similar to a phonogram, contains information about the extent of tool wear and the quality of the processed surface of the workpiece, which is inextricably linked to the tool wear. Studying the possibility of using a sound signal as an information signal began with determining its noise resistance—one of the main requirements for this kind of signals in the adaptive control of machining by cutting.

A cutting sound is recorded by microphone typically of electret type [46] (see Fig. 4.16). The voltage of the electrical signal at the output of the microphone varies according to a change in the recorded sound.

It is necessary to estimate the noise resistance of the cutting sound to the influence of noise disturbance, which is additive and multiplicative by nature. The first act in the measuring circuit is not critical for measuring computer networks, where such disturbance is practically absent.

In order to eliminate the multiplicative noise disturbance, it is necessary to present the measured microphone sound signal E_S in a dimensionless form. The dimensionless parameter is the ratio of the measured values of the sound signal E_{Si} to its value

Fig. 4.15 Surface of the
record magnified by a
microscope

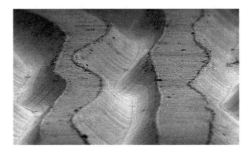

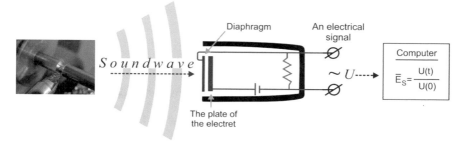

Fig. 4.16 Diagram of recording the cutting the sound by the microphone

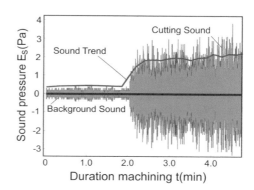

Fig. 4.17 Duration of cutting sound

E_{S0} recorded during the first measurement:

$$\overline{E}_S = \frac{E_{Si}}{E_{S0}}. \tag{4.13}$$

In the presence of multiplicative noise fraction, the Eq. (4.13) allows to exclude it as follows:

$$\overline{E}_S = \frac{E_{Si} \cdot \varepsilon(\tau)}{E_{S0} \cdot \varepsilon(\tau)} = \frac{E_{Si}}{E_{S0}}. \tag{4.14}$$

The cause of the additive noise is the background sound produced by the equipment surrounding this processing system. In the presence of additive noise fraction, the Eq. (4.13) takes the following form:

$$\overline{E}_S = \frac{\sqrt{\hat{E}_{Si}^2 + \hat{\varepsilon}^2(\tau)}}{\sqrt{\hat{E}_{S0}^2 + \hat{\varepsilon}^2(\tau)}} = \frac{\hat{E}_{Si}}{\hat{E}_{S0}} \frac{\sqrt{1 + \hat{\varepsilon}^2(\tau)/\hat{E}_{Si}^2}}{\sqrt{1 + \hat{\varepsilon}^2(\tau)/\hat{E}_{S0}^2}} \tag{4.15}$$

$$\hat{E}_S = \frac{E_S}{\sqrt{2}}. \tag{4.16}$$

$$\hat{\varepsilon}(\tau)_S = \frac{\varepsilon(\tau)}{\sqrt{2}}. \tag{4.17}$$

From Eqs. (4.16) and (4.17), it follows that the expressions under the square root in the numerator and the denominator in the general case are not equal and that they are not subject to reduction. This means that in practice of cutting there may be a danger of distorting influence of the noise disturbance on the measurement result T_{CR}. For this reason, the distorting effect of additive noise has been experimentally studied. Thus, Fig. 4.16 shows the timing of realization of the sound signal, comprising a background portion and a portion of the work of the lathe processing system (machine VF (HAAS)) during the cutting process. The experimental value of the T_{CR} was obtained during the finishing processing of the workpiece from steel 12X18H10T by cutting insert P25 in following modes: $S = 330$ m min^{-1}, $f = 0.15$ mm rev^{-1} and $a = 1.0$ mm. The layout of machines and the distance between them correspond to the "rules of occupational safety and usability of machines" [17]. The extent of the impact from nearby equipment interference was evaluated in accordance with the "rule of six decibels" known in acoustics [47], according to which the change in the sound (sound pressure) is directly proportional to the distance from the sound source to its point of control.

In this case, the rms sound pressure $\hat{E}_S^{SUM}$, Pa, in the control point is determined from the following expression:

$$\hat{E}_S^{SUM} = \sqrt{\hat{E}_S^2 + \left(\frac{\hat{E}_S}{4}\right)^2 + \left(\frac{\hat{E}_S}{4}\right)^2} = \hat{E}_S\sqrt{\frac{18}{16}} = 1.06 \cdot \hat{E}_S. \tag{4.18}$$

In the calculation of Eq. (4.18), it was assumed that in the process of work three processing systems generate the sound of the same amplitude E_S, Pa. As you can see, noise disturbance from the surrounding equipment leads to an overestimation of the measurement results with respect to the true value of the controlled sound magnitude by only 6%.

Therefore, the microphone located in the immediate (centimetre) proximity of the controlled working cutting zone virtually eliminates the distorting effect of the additive noise. This is because sound interference is undergoing a significant attenuation, overcoming the distance measured in metres from the source of its origin to the controlled work area.

To the category of additive noise, we refer interference that has the pulse (shock) character, which appears due to irregular mechanical shocks occurring in the surrounding machining equipment environment. Pulse load in the time domain of its submission is characterized by two parameters: the level E_{Sh} and duration τ_{Sh} (see Fig. 4.18).

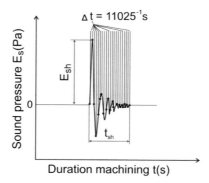

Fig. 4.18 Sound wave pulse and circuit of its digitization

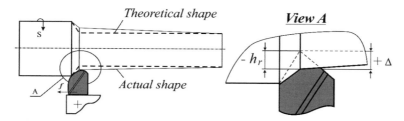

Fig. 4.19 Deterioration of the geometric accuracy of the part due to the amount of the wear of the tool blade

When the analogue signal, including pulse, comes into the computer, it is subjected to "digitization", i.e. discrete reading with a certain time step $\Delta \tau$. The time step $\Delta \tau$ is 1/11,025 s, where the number 11025 is the standard sample rate of discretization f_{dis}, used in the digital processing of analogue signals. The number of the read values of the signal n_{PCS} coincides with the number of time steps, which is determined from the following expression:

$$n_{PCS} = \frac{\tau_{Sh}}{\Delta \tau} = 11025 \cdot \tau_{Sh}. \tag{4.19}$$

The number of steps, as follows from Eq. (4.19), depends on the impulse frequency and duration pulse τ_{Sh}, which is determined by the following formula [6]:

$$\tau_{Sh} = \frac{2L}{C}. \tag{4.20}$$

Table 4.6 shows the impulse width and the corresponding number of the read values calculated, respectively, by Eqs. (4.5)–(4.6) for the length of the rod, varying from 0.1 to 1.0 m.

Table 4.6 Change of the pulse duration τ_{Sh} and the number of the read values of the sound impulse n_{PCS}, depending on the rod length L

L, m	0.2	0.3	0.4	0.5	0.6	0.7	0.8	0.9	1.0
$\tau_{Sh}\ 10^{-4}$, s	0.77	1.16	1.55	1.93	2.32	2.71	2.10	2.48	2.67
n_{PCS}	0.85 ≈ 1	1.27 ≈ 2	1.7 ≈ 2	2.13 ≈ 2	2.55 ≈ 3	2.98 ≈ 3	2.4 ≈ 4	2.8 ≈ 4	4

We estimate the contribution of the sound impulse in the recorded during the measurement of the total sound value. Suppose that in the absence of pulse interference, the amplitude of the useful sound signal is E_S, Pa. If there is interference of the total value of the amplitude, E_S^{SUM}, Pa is a vector sum of the interference level E_{Sh} and the level of the useful signal E_S. Assume that the noise level exceeds ten times, the useful signal level $E_{Sh}=10\ E_S$. Let us take a substantial margin that the pulse duration τ_{Sh} is such that it accounts for up to 100 time samples (it follows from the table that they are smaller by almost two orders of magnitude). In the process of the sound control, in order to improve the reliability of the measurement and control of the noise interference, altogether 3072 values of the signal are determined (reading is performed three times in 1024 samples). The read values are added and averaged by dividing the resulting sum by the number of terms—3072 units. For the determination of the unit vector of the useful signal and the amount of noise, we pass from the amplitude values of pressures in their rms values and determine the overall level of the useful signal and interference, Pa, then.

$$\hat{E}_S^{SUM} = \sqrt{\left(\hat{E}_S \frac{2972}{3072}\right)^2 + \left(10\hat{E}_S \frac{100}{3072}\right)^2} = 1,02\hat{E}_S. \qquad (4.21)$$

As it can be seen, even at tenfold excess of noise interference over the useful signal, its contribution to the total signal does not exceed 2%.

Thus, the results of the above studies revealed that the cutting sound has the properties of noise resistance, consequently, it can be considered, as the initial information signal used in the control process of cutting. At the same time, the cutting sound should ensure solving the basic problem of adaptive control—maintaining the quality required in the part manufacturing document, characterizing the degree of compliance of the part geometry and the purity of its surface with the requirements of the drawing documentation. Geometric accuracy is determined by the tool wear and surface finish by the magnitude of its roughness.

One of the reasons for deviation of the part geometry from the details of the drawing is the amount of the wear of the cutting tool h_r, equal in magnitude and opposite in sign of the radius change Δ of the processed workpiece surface (see Fig. 4.18).

For determining the correlation between the sound and tool wear during machining of the workpiece on a lathe, a special experiment was conducted. In the course of the experiments, the sound signal was measured continuously. Measurements were

Fig. 4.20 Microphone
placement in the immediate
vicinity of the cutting area

carried out via a microphone installed near the cutting zone (see Fig. 4.20) with the transmission of a signal to the computer. Simultaneously, in steps (in two tool passes) using a measuring digital microscope, we recorded the value of chamfer wear of the major back surface *VB*, in mm of the tool. The experiment was stopped when the maximum allowable value of the wear was reached.

The microphone was mounted on the cutter holder at a distance of 10 mm from the cutting zone.

The relation between the curve of the tool wear and the sound trend accompanying the machining by cutting using the insert P25 made of steel 12X18H10T for the following modes (330 m/min, $f = 0.15$ mm/rev, $a = 1.0$ mm) are shown in Fig. 4.21. Sample diagram of cutting sound change (parameter trend $\overline{E}_S$) and the curve of wear *VB* is shown in Fig. 4.21a, and information about the mutual correlation dependence between them is shown in Fig. 4.21b, characterized by the correlation coefficient value *R*, equal to 0.926 [20].

As we see, the correlation of cutting sound and tool wear is big enough, which objectively indicates a high degree of correlation of the change in the sound trend and the wear curve.

The roughness was measured at the longitudinal turning of steel billet with T15K6 on the modes presented above. The *Ra* parameter measurements were carried out

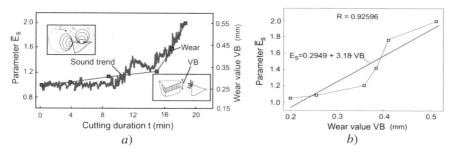

Fig. 4.21 a The curve of wear *VB* and the trend of sound $\overline{E}_S$; **b** the regression relationship between the dimensionless quantity of sound $\overline{E}_S$ and wear *VB*.

periodically every five passes using a cutter-type profiler. The results of the measurements were recorded by the instrument dial indicator and additionally recorded on a laptop. The signal recorded on a laptop was subjected to further processing in order to determine the correlation between:

- the sound trends and roughness parameter;
- the frequency spectra of the roughness profile and sound;
- the realization of a sound signal in time and the roughness profile.

The experimental results are shown in Figs. 4.22, 4.23, 4.24, 4.25. In addition, for clarity and ease of comparison, the roughness and sound, Ra–parameter, as well as the sound parameter $\overline{E}_S$, are given in a dimensionless form $\overline{Ra}$ $\left(\overline{Ra} = \frac{Ra(\tau)}{Ra(\tau_0)}\right)$.

Experiments have established as follows:

- The correlation coefficient R between the sound trends and roughness parameter Ra equals to 0.944 (see Fig. 4.22).

Fig. 4.22 Comparison of sound trends (parameter $\overline{E}_S$) and roughness ($\overline{Ra}$)

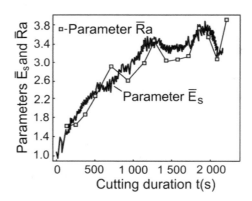

Fig. 4.23 Spectra of roughness $\overline{Ra}$ parameters and sound $\overline{E}_S$

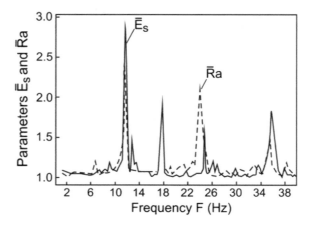

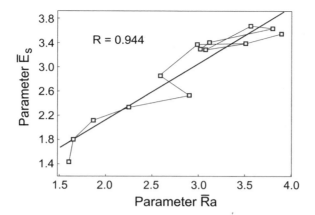

Fig. 4.24 Regression dependence between roughness $\overline{Ra}$ and sound $\overline{E}_S$ (parameters)

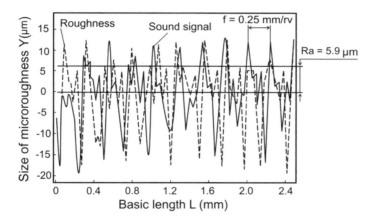

Fig. 4.25 Roughness profile and form of the sound wave

- The correlation coefficient R between the frequency spectra of the roughness profile and the sound is 0.783 (see Figs. 4.23, 4.24, 4.25).
- The coincidence, both on the level, and in turns of the peaks of the profile roughness and time of the sound realization (see Fig. 4.25, the y-axis, in this case, is scaled in absolute values in microroughness Y, μm; the figure also indicates the feed rate f and the value of Ra roughness, measured by a profilometer).

Thus, the results of experiments indicate a close correlation of sound generated by the cutting process with tool wear and the quality of the treated surface. The novelty of the material discussed in the chapter is that for the first time a close correlation relationship between the cutting sound and the degree of tool wear and roughness of the treated surface is shown.

The results of the studies indicate the expediency of using a sound signal as initial information in the adaptive control of machining materials by cutting. The results are based on the following experimentally proved provisions:

- in contrast to the data signals, measured via contact method and subject, there-fore, to noise disturbance generated by mechanical vibrations of the machine, a non-contact method of sound measurements via a microphone placed near the cutting area, and the transition in the processing of the measurement results to the dimensionless quantity of the sound $\overline{E}_S$, provides noise resistant testing results to multiplicative and additive by nature noise interference, as well as interference of the pulse character;
- sound trend and tool wear curve change during the cutting process with a high degree of, correlation as proven by a significant value of their mutual correlation coefficient $R = 0,926$.
- Strong correlation was established between:
- the sound trends and altitude roughness parameter Ra, characterized by a correlation coefficient R, equal to $0,944$;
- the frequency of audio spectra and roughness profile characterized by a correlation coefficient R, equal to $0,782$.

Thus, the data given in the chapter allow us to recommend the sound for effective control of the materials processing by cutting and forecast the durability of the cutting tool.

4.5.3 Adaptive Control of Cutting Conditions Based on Individual Resource Forecast of the Cutting Tool

4.5.3.1 General Statements

Adaptability of cutting-edge machining performed on CNC machines is ensured by the choice of optimal combinations of three controllable parameters according to one or another algorithm: depth, feed and cutting speed. Depending on the problem being solved, the optimum is provided by determining the extremum of the objective function, which allows us to achieve either a maximum of performance or a minimum of the cost of the part produced, or the maintenance of the limit value of the adjustable parameters [26, 27].

Even if the adopted algorithms of adaptive control are well developed, a significant problem of cutting metals is the problem of effective control of the cutting tool condition, which has not been solved yet [28]. Currently, progressive methods of direct or indirect monitoring of the tool condition are limited only by comparing of actual wear rate of the tool with its standard rate determined by one or another, sometimes quite sophisticated method [29].

However, this does not set the task of forecasting the tool running time before it reaches its maximum allowable wear rate, which makes it impossible to make a decision in advance about changing the machining mode in order to prevent tool sudden failure and thereby eliminating the appearance of a defective part. The absence of such a forecast negates all efforts to develop an adaptive control algorithm, no matter how effective, in the opinion of its authors, it is.

The lack of effective methods for forecasting the tool resource is partly explained by the fact that even knowledge of the actual wear rate does not allow with sufficient probability to forecast the tool running time before reaching the maximum allowable wear rate. This is explained by the fact that normative wear rate is purely a statistical average. For this reason, the wear rates characterize the actual conditions of tool loading only with a certain degree of reliability determined by a specific combination of cutting conditions and materials of both the tool and the workpiece.

It is not possible to normalize the tool resource (durability) resulting from its loading conditions. This is explained by the fact that in practice there is an infinite number of actual tool loading conditions that are not amenable to normalization. Data on the tool resource as well as wear rates, provided by a manufacturer, are of average statistical character with a small degree of probability, characterizing the individual tool resource, determined by specific development of conditions of its operation.

As a rule, the result of this uncertainty is a sudden failure of the tool, which in turn causes an incorrigible defect of the part.

The solution to this problem was found in the use of the new methodology for forecasting the individual resource of mechanical systems considered in the monograph. The presented methodology has formed the algorithmic basis of the automated system for adaptive control of the cutting-edge machining.

The novelty of the tool resource forecasting methodology, unlike current methods focusing on average data on wear limit and tool resource, consists in real-time forecast of the individual tool resource meeting actual conditions of its loading.

The development, which is based on the presented principles of the automated system for the cutting process adaptive control, is undoubtedly a topical scientific and practical task.

The subject of the study was the adaptive control system based on the real-time forecast of the individual resource of the cutting tool.

The aim of the work, the main results of which are presented in this chapter, was the development of the algorithm and the corresponding software forming the basis of the automated system of the cutting process adaptive control that ensures implementation of the new methodology of forecasting individual resource of the cutting tool into engineering practice.

4.5.3.2 Algorithm of Adaptive Control of the Cutting Process

In the process of metalworking systems control, we have to deal with the situation when the mathematical model of the control object is not fully known, and dynamic characteristics of the object, described by this model, are changing continuously depending on internal conditions influenced by the change of the workpiece geometry and the contour of the cutting tool edge due to its wear during the cutting process [29].

In this case, it is proposed to consider a mathematical description of the trend of information signals different physical nature, which are accompanying the cutting process, as a model of the control object. Sound, generated at the contact point of the workpiece and the tool, is the most informative signal [30].

The model, describing the behaviour of the sound trend, considered as the model of dynamic behaviour of the machining system, made it possible to develop and propose a completely new methodology for forecasting the cutting tool resource for wide use in actual production, which formed the basis for the automated system for cutting process adaptive control. The control algorithm, developed in accordance with this methodology, combines the solution of identification and control problems within a single process.

Based on the results of identifying the trend model in real time, an individual tool resource is forecasted. Further, on the basis of comparison of the forecast with the required machining time for those processes, the choice of the optimal cutting mode is selected, which makes it possible to prolong the period of the tool defect-free operation reducing to almost zero the probability of occurrence of a defect in the part produced.

When developing the control algorithm, the two well-known facts were taken into account. The first: a decisive influence on the quality of the surface being machined is the cutting tool edge wear, in particular, its back main surface wear VB. The second: duration of defect-free machining of the part is determined by the individual tool resource T_{IND} realized in the given conditions of its loading. The maximum allowable wear rate $[VB]$ and the resource T_{IND} are connected by the following relationship [31]:

$$T_{IND} = \frac{[VB]}{S \cdot \gamma}. \tag{4.22}$$

The value $[VB]$ is standardized ($[VB] = $ const); therefore, duration of the tool defect-free operation can be controlled (to control its resource T_{IND}), as it follows from the relationship (4.22), only by changing the cutting mode (cutting speed S) and wear intensity γ. The purpose of varying of these parameters is to meet the following condition during the cutting process:

$$T_{IND} \geq T_{REQ} \tag{4.23}$$

The required machining time T_{REQ} is calculated using the following formula [32]:

$$T_{REQ} = \frac{\pi D_D \cdot L_D}{1000 \cdot S_0 \cdot f_0} \tag{4.24}$$

The rate of the individual tool resource T_{IND} is determined in the process of parametric identification of the sound trend model (3), carried out according to the results of real-time measurement of the sound level E_S^f accompanying the cutting process [30]. Identification consists in minimizing the discrepancy (4) of the calculated E_S^C and actual values of the trend (time series) E_S^f of the sound.

The wear intensity γ is proportional to the pressing force production P_{PR}, acting in the friction pair and the velocity of their relative slip V_{SLIP} [31]. In this case, the slip velocity V_{SLIP} is equal to the cutting speed S, and the pressing force is equal to the cutting force P_{CUT}; therefore, as applied to the cutting process, this condition is written as follows:

$$\gamma \sim P_{CUT} \cdot S. \tag{4.25}$$

In this case, the friction pair is the tool working surfaces and the workpiece cutting surface. The cutting force is determined by the following formula [32]:

$$P_{CUT} = \sqrt{P_x^2 + P_y^2 + P_z^2}. \tag{4.26}$$

The components of the cutting force $P_{x,y,z}$ are determined by the empirical power law dependence of the form [34]:

$$P_{x,y,z} = 10 \cdot C_p a^x f^y S^n K_p. \tag{4.27}$$

The coefficients values and degrees exponents contained in (4.27) are given in the corresponding directories [17].

Dividing both sides of the inequality (4.23) into processing time required by the technical process T_{REQ}, taking into account expressions (4.22), (4.25), (4.27) and meeting the condition $a = const$, we get the following

$$\frac{T_{IND}}{T_{REQ}} = \frac{S(\tau)}{S_0} \cdot \frac{\gamma(\tau)}{\gamma_0} = \frac{S(\tau)}{S_0} \cdot \frac{P(\tau) S(\tau)}{P_0 S_0}. \tag{4.28}$$

Having subtracted from the left side of the equality (4.28) its right side, and having squared the resulting expression, we get the expression for the objective function:

$$U = \left[\frac{T_{IND}}{T_{REQ}} - \left(\frac{S(\tau)}{S_0} \right)^2 \cdot \frac{P(\tau)}{P_0} \right]^2. \tag{4.29}$$

The values of the varying cutting mode parameters $S(\tau)$ and $P(\tau)$ are selected from the condition of the inequality (4.23). In this case, the range of varying cutting modes is limited by the degree of change in cutting force. Given this limitation, the objective function (4.29) takes the following form:

$$U = \left[\frac{T_{IND}}{T_{REQ}} - \left(\frac{S(\tau)}{S_0} \right)^2 \cdot \frac{P(\tau)}{P_0} \right]^2 + \left[1 - \frac{P(\tau)}{P_0} \right]^2. \tag{4.30}$$

Adaptive control of edge cutting mode is carried out according to the following algorithm:

- According to the procedure adopted in metal cutting technology [17] and depending on the adaptive system class (optimization system or boundary control), the initial processing mode is assigned, i.e. the initial values of the control parameters are set: a_0, S_0, f_0.
- In real time, the sound level E_S^f generated by the cutting process is measured, and based on the results of these measurements, the trend model is identified, and the determined model parameters include the numerical value of the required tool resource T_{IND}.
- If the tool resource exceeds the required time to machine the part (s), i.e. the condition $T_{IND} > T_{REQ}$ is satisfied, then the two possible variants are possible.
- If the cutting process is carried out in a single or small-scale production of expensive products, then it continues within the initial mode.
- If the cutting process is carried out within the conditions of mass production of low-cost parts, it is possible to force (tighten) cutting modes in order to increase productivity (the decision is made by the operator).
- If the relationship $T_{IND} < T_{REQ}$ is observed, then for the both types of production, the cutting mode is changed to fulfil the condition (4.23).

Moreover, when machining long objects, the required resource T_{REQ} is assumed to be equal to the time required to complete the technological passage T_{PAS} $(T_{REQ} = T_{PAS})$.

$$T_{PAS} = \pi D_D \cdot \frac{L_D}{1000 \cdot S(\tau) \cdot f(\tau)}. \tag{4.31}$$

To automate the system of cutting process adaptive control, which reproduces the given algorithm based on the new methodology for forecasting the individual resource of the cutting tool, a special hardware and software system has been developed to complete the standard system of a CNC machine.

4.5.3.3 Hardware and Software System of the Cutting Process Adaptive Control

In metalworking, the level of optimality of the adopted cutting conditions depends on how accurately the initial information characterizes the actual conditions of the machining process. In other words, to what extend the following characteristics are being changed: allowance, hardness of the material being machined, machining system rigidity, tool wear and its resource adopted in the process of calculation and preparation of the technological process and control program [7, 26].

Currently, the operation of majority of automated machines is subject to "hard" programming focusing on average statistical data (norms) of tool wear and its durability [35–43]. For example, depending on the quality of the cutting tool, its resource in one batch varies from 15 to 35% of average value. If the tool operation time is determined by the worst sample in a batch, then the most stable samples, with a fixed operating time, use their resource only by 65% at best [29].

In this case, a resource in accordance with [34] means a time resource equal to the operating time of the cutting tool from the start of cutting with a new tool until it reaches the limit state.

Due to the instability of the tool life, the main part of the cutting process proceeds either with the underutilization of the cutting tool capabilities or does not exclude the appearance of a part failure due to unforeseen tool failure [34]. To eliminate these failures, a forced replacement of the tool is envisaged, and as a result, it is removed from service prematurely having substantial resource reserve [35].

Monitoring of the tool condition directly during the cutting process in order to determine the degree of its criticality as well as to forecast its operating time before replacement is a rather complicated process. For example, measuring of blade cutting tool wear should be carried out after each machining cycle [44], but in this case it is difficult to forecast the onset of a critical tool condition: ultimate wear, chipping, breakage, and, moreover, to make long-term forecasts to determine the moment of its replacement.

Such control requires, as a rule, the interruption of the machining process, which significantly reduces its effectiveness. Therefore, indirect methods are often used for online diagnostics as they control the tool condition according to the sound level accompanying the cutting process [40].

A common disadvantage of the existing forecasting methods based by their very nature on standard normative values is that they lead to significant forecast errors, which makes it impossible to use them in adaptive control of the cutting process.

The algorithm of adaptive control presented in Chapter 4.5.2.3 served as the basis for the development of the software and hardware system (Fig. 4.26).

The software part of the system reflects the above-described new methodology for forecasting the resource of mechanical systems in relation to the cutting tool. This provides a choice of optimal cutting conditions, which allow to extend the period of defect-free machining of the workpiece.

The software product is built according to the modular principle, which allows to change flexibly its structure in relation to the features of the cutting mode, the properties of the workpiece materials and tools.

To ensure the versatility of the software system in relation to the hardware (Fig. 4.27), it has been implemented in several algorithmic languages: Turbo Pascal, Delphi, C and JAVA. The latest version of the system is designed for microprocessor-based devices running on the Android platform.

The system provides registration, accumulation, processing of information and adaptive control based on monitored equipment operation. In accordance with the above-described algorithm for adaptive control of the cutting process, the operation of the software and hardware system (Fig. 4.26) begins with the initial data input: cutting modes and geometrical parameters of the workpiece and the part, on the basis of which the required duration of the processing system (T_{REQ}) is calculated.

Fig. 4.26 Schematic block diagram of the software and hardware system

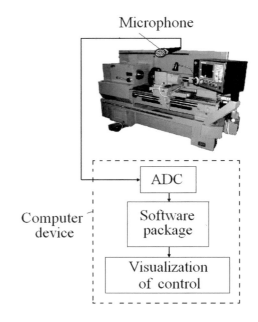

Microphone

Computer device

ADC

Software package

Visualization of control

Next, the amplitude of the sound wave E_S is recorded and the individual resource (T_{CR}) of the instrument is forecasted. If the forecasted value of the resource exceeds the required $(T_{CR} > T_{REQ})$, then machining of the part continues in the previous modes. Otherwise, the machine is transferred to a new automatically calculated mode of its operation, ensuring the fulfilment of this condition.

Below is an example of adaptive control of a CNC turning machine 16К20Т1 based on the presented algorithm. The kinematic scheme of the machine allows us to change the spindle rotation n in the third mode of its operation from 125 to 2000 rpm, and the feed f can be changed with spacing of 0.01 mm/rev from 0.01 to 2.8 mm/rev.

The workpiece was subjected to semi-finishing cutting with a cutting plate made of hard alloy T15K6 in the following initial conditions: $S_0 = 63$ m/min, ($n = 100$ rpm), $f_0 = 0.5$ mm/rev, $a = 1.0$ mm. Modes were selected according to the requirements of the directory [17]. The length of the part was 310 mm, the number of passes was three, the diameter of the workpiece was 200 mm, and the diameter of the part was 194 mm. The exponents in the formula (4.27) had the following values: $n_{Px} = -0.4$ (for P_x); $n_{Py} = -0.3$ (for P_y), $n_{Pz} = -0.15$ (for P_z); $y_{Px} = 0.2$ (for P_x), $y_{Py} = 0.8$ (for P_y), $y_{Pz} = 0.9$ (for P_z). The machining time required by the technical process (machine time) T_{REQ} was 18.6 min.

The sound was monitored continuously throughout the entire machining process using a microphone placed near the cutting zone [15]. The signal from the microphone (Fig. 4.20) was sent to the "sound card" of the computer, where it was digitized and further processed. According to the above algorithm for adaptive control of the cutting process based on the results of sound signal processing, an individual tool resource T_{CR} was forecasted, and by minimizing the objective function (4.30), the optimal

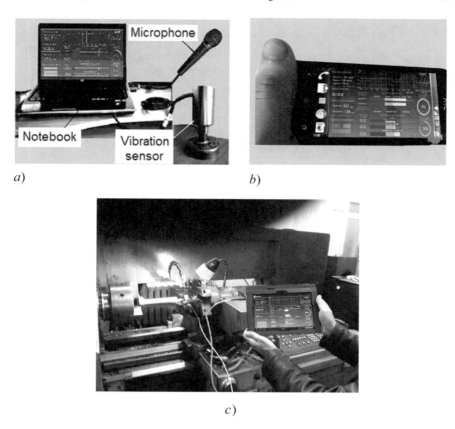

Fig. 4.27 Options for implementation of the forecast diagnostic hardware complex: **a** laptop-based; **b** smartphone-based; **c** tablet-based

combination of controlled parameters (cutting speed $S(\tau)$ and longitudinal feed $f(\tau)$) was chosen, ensuring compliance with the condition (4.23).

The results of the experiment are presented in Fig. 4.28 and 4.29. Figure 4.29 shows the relationship between the machining time required for a part T_{REQ} and the forecast of an individual tool resource T_{CR} corresponding to the tool loading at the initially accepted cutting modes (S_0 and f_0).

Figure 4.29 shows the actually realized cutting conditions ($f(\tau)$ and $S(\tau)$), which made it possible to complete the machining of the part without changing the tool and thereby to ensure the fulfilment of the condition (4.23) during the cutting process. Actually realized duration of defect-free processing (the corrected value of the tool resource) T_{ACT} is shown in Fig. 4.29.

Thus, the experiment showed that during the part machining, the forecast of the individual resource T_{CR} did not exceed the machining time T_{REQ} required by the technological process (Fig. 4.28), i.e. the condition (4.23) was not satisfied. At the

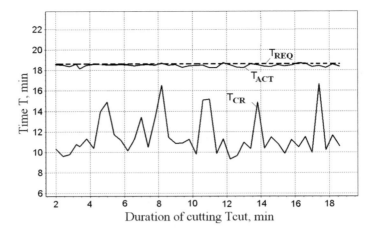

Fig. 4.28 Comparison of the required cutting duration T_{REQ} with individual resource forecast T_{CR} and its adjusted value T_{ACT}

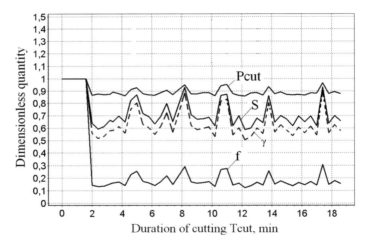

Fig. 4.29 Change in the process of adaptive control of regulated parameters $f(\tau)$, $S(\tau)$ as well as cutting forces P_{CUT} and wear intensity γ

same time, the individual resource was on average equal to 12 min, which amounted to about 67% of the required machining time. To fulfil the condition (4.23), it was necessary to reduce the wear rate of the tool by 33% (Fig. 4.29). The decrease in wear intensity was achieved by reducing the cutting speed S by 30% and feed f was reduced by 80% on average.

The change in cutting conditions led to a decrease in the cutting force P_{CUT} by an average of 10% (Fig. 4.29), resulting in a decrease in the tool load, which, along with a decrease in the cutting and feed speed, prolonged its resource and, consequently, the defect-free machining period the part.

The experience of using the presented adaptive control methodology shows that, as a rule, the feed f undergoes the greatest changes having the most significant impact on a tool wear intensity (intensity of its loading) at a constant depth of cut a.

4.6 Forecasting in Medicine

The methodology of forecasting events and phenomena considered in this monograph as applied to medicine makes it possible to determine the moment of occurrence of the limit of working capacity of each of the employees individually.

This assessment is performed on the basis of the spectral analysis of electrocardiogram records, separation of the spectral component that coincides with the heart rate, цреатион of a time series of amplitudes of this component and forecasting based on examination of the nature of changes in this series in the work process of a person till the moment of his/her exhaustion, which is the novelty of this research. To solve this problem, it is necessary to carry out proactive forecasting of the onset of this event [48–52].

This circumstance has served as a serious stimulus for the development of a new forecasting methodology, called "cardio forecasting", which is radically different from the existing ones [53–57].

Traditionally, forecasting consists of two stages, the first of which, based on available data, determines the parameters of the approximating function, and the second one determines extrapolation of the graph of this function to the intersection with the maximum allowable level of the monitored parameter. The coordinate of the point of intersection on the x-axis is the desired argument of the function, which determines the moment when the controlled event or phenomenon reaches its maximum permissible (critical) state [15].

However, it should be noted that this forecast, unfortunately, is not reliable enough. This is explained by the fact that the maximum permissible criterion value of the monitored parameter is of a moderately statistical nature. As a result, the average statistical value of the monitored parameter describes only with a certain degree of confidence the actual state of the given person. This circumstance leads to errors in forecasting the moment when a person reaches the (individual) fatigue limit that is characteristic only for him/her [15].

Also, the existing methods of forecasting the state of a human operator, despite their external diversity, are guided by comparing the current (static, frozen) physiological portrait of a person with certain his/her etalon. But, strictly speaking, there is no such reference general etalon for a person.

Figuratively speaking, the etalon is "inside" a person. The operator, in a certain sense, is an etalon for himself/herself.

The search and implementation into a practice of the methodology for assessing and forecasting changes in the functional state of a human operator in the process of performing professional activities are an actual scientific and practical task. Examples of applying this methodology are given below.

4.6.1 Subject of Research and Research Procedure

The subject of research was the recordings of cardiograms of four operators that were taken during the work shift with a duration of 14 h (operators 1 and 2) and within two working weeks with a break for rest (operators 3 and 4).

The research procedure, carried out to confirm the effectiveness of the use of "cardio forecasting" for determination of the moment of achievement of fatigue limit by each of the individuals in the given conditions of work that were specific only to him/her, assumes monitoring of the performance capability of a human operator in the "online" mode during working hours.

The studies investigated two types of forecasting—short term and long term. Short-term forecasting provides a forecast for one or two work shifts, while the long-term forecasting—for one or two working weeks.

Electrocardiograms, which were recorded continuously during the working time, excluding the rest period, were used as the initial source data (see Fig. 4.30).

Figure 4.30a shows the typical cardiogram records and their spectra, used as source input data for forecasting the degree of working capacity of a human operator. In Fig. 4.30b, the HR sign indicates the heart rate, the amplitude of which is a symptom of the degree of criticality of the state of the operator (worker).

The recording of pulse oscillations was performed, using a fitness bracelet, information from which was transmitted via the Bluetooth system to a smartphone using the Android OS. The software installed in the smartphone is subjected to the recorded signal to spectral analysis. The sampling rate was 1024 Hz, and the number of the read points was 8192 pieces, which provided a frequency step of 0.125 Hz. The results of this processing are stored in the "Internet cloud" under the name of a human operator, creating thus a "database" used for forecasting of his/her working capacity.

The frequency component was singled out from the spectrum that corresponded to the heart rate (HR). According to the results of the regular measurement of this amplitude, a time series was compiled (a trend of the amplitude of HR). This series was approximated by a trend model in order to determine its coefficients.

From the point of view of mathematics, the approximation consists in minimizing the difference between the m pairs of calculated values A_{HR}^C and actual values A_{HR}^{ACT} of the HR amplitude, which characterized the person's fatigue at each of the m moments of control of his/her state (model (3)). The minimization of the functional (4) was performed for forecasting T_{CR} duration of work of the human operator till fatigue.

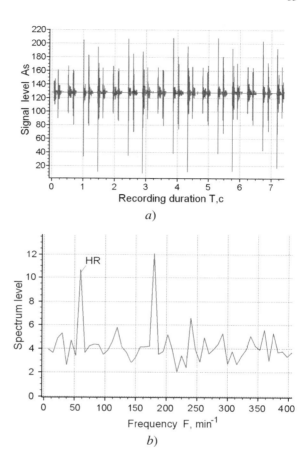

Fig. 4.30 Source data for forecasting: **a** temporary recording of a signal (cardiogram); **b** heart beat spectrum

The quality (plausibility) of the forecast was assessed by the results of its comparison with the length of working time. The stability of the short-term forecast was also considered. The degree of its variability was estimated by the standard deviation of the performance forecasting (rms).

The quality (probability) of the forecast was assessed by the results of its comparison with the length of working time. The stability of the short-term forecast was also considered. The degree of its variability was estimated by the value of the mean square deviation of the forecasting of the working capacity (rms).

This chapter considered two types of forecasts: short term and long term; the first examines a category of workers working for two shifts in a row (cargo drivers for long distances, supermarket workers, etc.), and the second one examines working by keeping watch, etc.

4.6.2   *Results of Research and Their Assessment*

4.6.2.1   Short-Term Forecasting

The results of short-term forecasting are presented in Fig. 4.31, 4.32, 4.33 and 4.34.

As it follows from Fig. 4.31 and 4.33, the forecasting makes it possible already at the beginning of the working period (from the first to the second hour of work) to consistently forecast the time of exhaustion of human working capacity. Moreover, as we see (see Figs. 4.32 and 4.34), the deviation of the forecast from the actual time of exhaustion of working capacity is within 30 min. The variability of the forecast is also insignificant, and it is characterized by the value of rms, varying within 20–22 min(see Figs. 4.31b, 4.33b).

Fig. 4.31 Forecasting results for operator 1: **a** approximation of the source data by the trend model; **b** the results of the forecasting

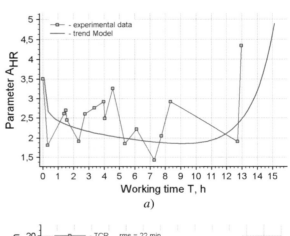

a)

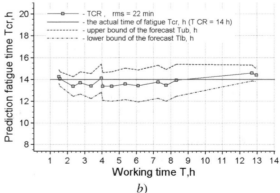

b)

Fig. 4.32 Deviation of forecasting of working capacity from the actual time of its exhaustion for the operator 1

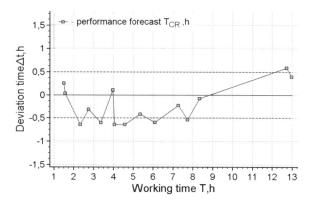

Fig. 4.33 Forecasting results for operator 2: **a** approximation of the source data by the trend model; **b** the results of the forecasting

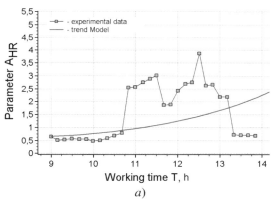

a)

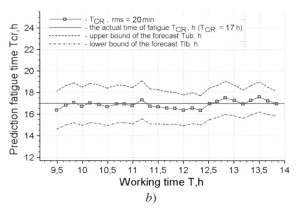

b)

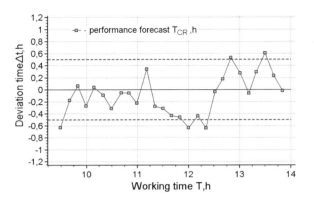

Fig. 4.34 Deviation of forecasting of working capacity from the actual time of its exhaustion for the operator 2

4.6.2.2   Long-Term Forecasting

The results of long-term forecasting are presented on Figs. 4.35, 4.36, 4.37, 4.38.

As it follows from Figs. 4.35 and 4.37, the use of the forecasting for long-term fore-casting makes it possible, already at the beginning of the working period, measured in days, to forecast the moment when the working capacity of the human operator is exhausted. At the same time, the deviation of the forecast from the actual time of exhaustion of working capacity is within half a day (Figs. 4.36 and 4.37).

The variability of the forecast is also insignificant, and it is characterized by a rms value not exceeding 7.6 h(see Figs. 4.36b and 4.37b).

Thus, the above results of forecasting clearly show its advantages compared in comparison with the current method of forecasting the state of working capacity of a human operator that is oriented on the etalon of working capacity of a human operator.

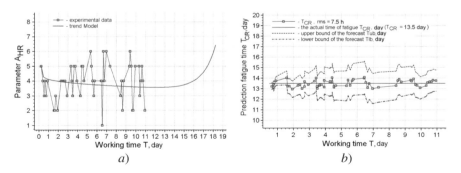

Fig. 4.35 Forecasting results for operator 3: **a** approximation of the source data by the trend model; **b** the results of the forecasting

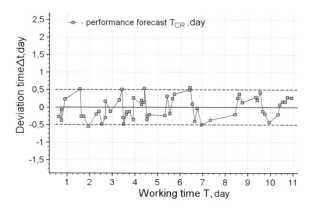

Fig. 4.36 Deviation of forecasting of working capacity from the actual time of its exhaustion for the operator 3

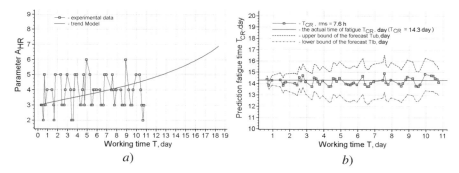

Fig. 4.37 Forecasting results for operator 4: **a** approximation of the source data by the trend model; **b** the results of the forecasting

Fig. 4.38 Deviation of forecasting of working capacity from the actual time of its exhaustion for the operator 4

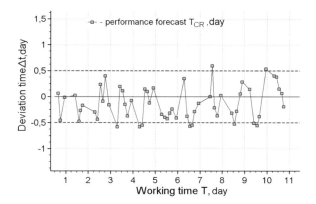

4.7 Earthquake Forecasting

4.7.1 Problem Statement

Earthquake forecasting is based on the different signal's registration, such as geophysical, geochemical, mechanical, geological, meteorological and other earthquake precursors. Their number exceeds three hundred [58]. This fact indicates that each of them does not fully meet the following requirements for them [59]:

- clear physical meaning of prognostic features;
- the physical validity of the connection of each prognostic feature with the preparing earthquakes process;
- provide each prognostic features with observational data, both in time—the presence of long-term series of prognostic features values, and in space, i.e. the possibility of their mapping;
- the presence of a formalized procedure for identifying anomalies of prognostic features based on a model of their behaviour during the earthquake preparation period;
- the possibility of obtaining estimates of the retrospective statistical characteristics of each precursor: successful forecasts probabilities (detection probabilities), false alarm probabilities, forecast efficiency (informativeness), etc.

At the same time, generally accepted methods for forecasting earthquakes, regardless of the physical nature of their precursors, in line with the current approach to forecasting in general, ultimately boil down to comparing the current seismic situation with some kind of its standard. However, the lack of such standards in the forecasting seismic phenomena practice significantly limits the effectiveness of these techniques.

The earthquake forecasting method is devoid of these shortcomings and is based on the trend analysis of the changes in the precursor value during the earthquake preparation period.

It should be noted, that irrespective of the physical nature of one or another earthquake precursor, the behaviour of its trend during the earthquake preparation period is in itself a very informative prognostic feature.

4.7.2 Initial Data

According to the methodology, considered in this chapter, the results of seismic situation continuous monitoring in the country are carried out through its national network of seismic stations, used as raw data in earthquake forecasting, as an example of such seismic stations network, used the F-net network [60] located in Japan (See Fig. 4.39).

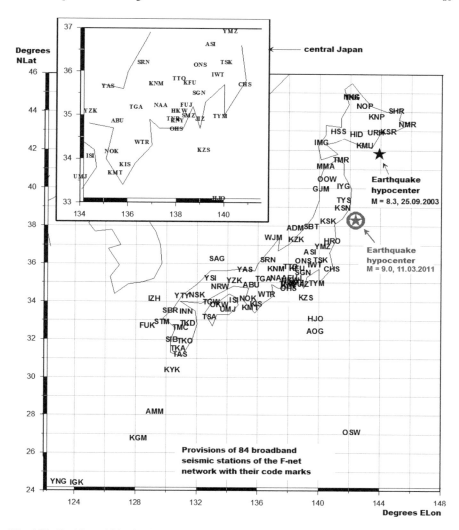

Fig. 4.39 Position of 84 seismic stations of the Japanese network F-net

A typical form of the recorded seismic signals representing regular time series is shown in Fig. 4.40.

As you can see, the time signal is amplitude modulated. The frequency of the modulating signal is due to the lunisolar tidal effect on the Earth. The period of this exposure is 13.75 days [61].

According to the monitoring results, a zone with an increased seismic vibration of the earth's surface is detected. Subsequently, the measurements performed by each of the seismic stations located in this area (hereinafter referred to as reference seismic

Fig. 4.40 Example of
recording the temporal
realization of a seismic
signal

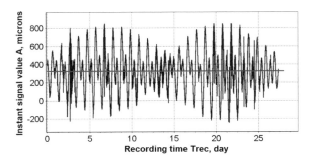

stations) are subjected to a more detailed analysis in order to prepare initial data for
forecasting earthquake.

Source data list includes the same:

– number (N) of reference seismic stations and geodetic coordinates x, y their
  location;
– calendar time t_j measurements taken at each of the reference stations.

4.7.3 Methodology for Preparing Initial Data for Forecasting

Initial data preparation provide.

– The seismic signal spectral analysis is recorded by each of the N reference seismic
  stations, highlighting the infra-low-frequency part of the spectrum (See Fig. 4.41)
  containing frequency components.

The spectrum contains frequency components of the spectrum located on:

Fig. 4.41 Example of a
typical infra-low-frequency
part of a seismic signal
spectrum

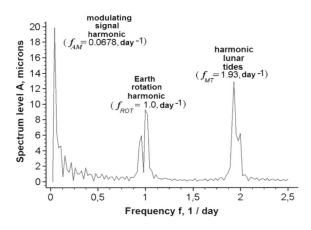

- Earth's rotation frequency 1.0, day^{-1};
- frequency of following the Moon tides $= 1.93$, day^{-1};
- frequency of the modulating signal $= 0.0678$, day^{-1}.

The periods of these components are equal [61]:

- $T_{AM} = 13.75$ day (amplitude modulation period f_{AM} ($f_{AM} = 1 / T_{AM} = 1/13.75$ $= 0.0678$ day^{-1}) seismic signal (Fig. 4.40, 4.41), due to the periodically repeating mutual arrangement of the Earth, the Moon and the Sun;
- $T_{ROT} = 1$ day (Earth rotation frequency period f_{ROT} ($f_{ROT} = 1 / T_{ROT} = 1/1 = 1$ day^{-1}));
- $T_{MT} = 12.42$ h (the period frequency of semi-daily Moon tides f_{MT} (hereinafter the frequency of Moon tides), $f_{MT} = 1 / T_{MT}{\cdot}24 = 24 / 12.42 = 1.93$ day^{-1}));
- the formation for each of the N reference seismic stations on the basis of the amplitudes A_{MT} frequency component located at the Moon tides frequency f_{MT} (hereinafter tidal harmonics), and the calendar time t_j of their registration, the trend $A_{MT}(t_j)$ characterizing the change in these amplitudes during the earthquake preparation period.

Based on the prepared initial data, the earthquake forecasting procedure is then performed.

4.7.4 Forecasting Method

Forecasting involves determining the moment of next earthquake occurrence, indicating its focus coordinates and strength. Consider the forecasting procedure in more detail.

4.7.4.1 The Forecasting of the Moment of the Occurrence Earthquake and Its epicentre's Coordinates

The earthquake time and its source coordinates are determined in the minimizing the functional U (4.32):

$$U = \sum_{j=1}^{m} \left[A_{SW}(t_j) - A_{MT}^{C}(t_j) \right]^2. \tag{4.32}$$

The tidal harmonic amplitude $A_{SW}(t_j)$ of a seismic wave at a point located at a unit distance ($L_1 = 1$ km, 1 m) from the maturing earthquake source is determined by the following formula:

$$A_{SW}(t_j) = \sqrt{\sum_{i=1}^{N} \left(\frac{A_{MTi}(t_j)}{L_i} \right)^2}. \tag{4.33}$$

The distance from the i-th reference seismic station to the maturing earthquake center [62], measured in km, is determined by the following formula:

$$L_i = arctg \left\{ \frac{\sqrt{(cosx_1 sin\Delta y_i)^2 + [cosx_i sinx_1 - sinx_i cosx_1 cos\Delta y_i]^2}}{sinx_i sinx_1 + cosx_i cosx_1 cos\Delta y_i} \right\} \cdot 6372.8. \tag{4.34}$$

The forecast model at a point located, at a unit distance from the maturing earthquake source, is determined by the following formula:

$$A_{MT}^{C}(t_i) = A_{SW}(t_0) \cdot \left[1 + \gamma \cdot \frac{(t_i - t_0)^p}{(T_{UB} - t_i)^q} \right]. \tag{4.35}$$

The tidal harmonic amplitude of the seismic signal at a point located at a unit distance from the maturing earthquake source that occurred at the initial time t_0 is determined by the following formula:

$$A_{SW}(t_0) = \sqrt{\sum_{i=1}^{N} \left(\frac{A_{MTi}(t_0)}{L_i} \right)^2}. \tag{4.36}$$

Substituting in (4.32) the formula for the seismic wave trend amplitude $A_{SW}(t_j)$ (4.33) and the forecast model $A_{MT}^{C}(t_j)$ (4.35), we move from the general expression (4.32) to the working formula (4.37), which together with the expression for the distance L_i (4.34) form a system of two Eqs. (4.38).

$$U = \sum_{j=1}^{M\Sigma} \left[\sqrt{\sum_{i=1}^{N} \left(\frac{A_{MT\,i}(t_j)}{L_i} \right)^2} - \sqrt{\sum_{i=1}^{N} \left(\frac{A_{MT\,i}(t_0)}{L_i} \right)^2} \cdot \left[1 + \alpha \frac{(t_j - t_0)^p}{(T_{UB} - t_j)^q} \right] \right]^2 \Rightarrow min. \tag{4.37}$$

$$\begin{cases} U = \sum_{j=1}^{m\Sigma} \left[\sqrt{\sum_{i=1}^{N} \left(\frac{A_{MTi}(t_j)}{L_i} \right)^2} - \sqrt{\sum_{i=1}^{N} \left(\frac{A_{MTi}(t_0)}{L_i} \right)^2} \cdot \left[1 + \alpha \frac{(t_j - t_0)^p}{(T_{UB} - t_j)^q} \right] \right]^2 \Rightarrow min, \\ L_i = arctg \left[\frac{\sqrt{(cos\,x_O\, sin\,\Delta y_i) + (cos\,x_i\, sin\,x_O - sin\,x_i\, cos\,x_O\, cos\,\Delta y_i)^2}}{sin\,x_i\, sin\,x_O + cos\,x_i\, cos\,x_O\, cos\,\Delta y_i} \right] \cdot 6372.8 \end{cases} \tag{4.38}$$

As a result of the equations system solution (4.38), five parameters are determined: T_{UB}, x_O, y_O, α, p, q, the main of which is the T_{UB} parameter and the geodetic coordinates of the x_O, y_O maturing earthquake source.

When solving this system, the main operation is minimization (4.32) of the forecast model $A_{MT}^C(t_j)$ (4.35) standard deviation from the trend of the seismic wave amplitude tidal harmonic $A_{SW}(t_j)$ (4.36).

Based on the parameter T_{UB}, which is the upper bound of the forecast range, the lower bound of the forecast range T_{LB} is determined from expression (4.39). In this case, the parameter expectation and the calendar time t_j, measured in year fractions, are considered as the lower bound:

$$T_{LB} = t_j \cdot \left(1 - P(t_j)\right) + T_{UB} \cdot P(t_j). \tag{4.39}$$

The probability of no earthquake at the current time $t_j P(t_j)$ is determined by the following formula:

$$P(t_j) = exp(-t_j / T_{UB}). \tag{4.40}$$

Further, based on the boundary values of the forecast range of T_{UB} and T_{LB}, their average value is calculated, taken as the most probable forecast time value T_{MPV} of the earthquake onset, measured in year fractions:

$$T_{MPV} = \frac{T_{UB} + T_{LB}}{2}. \tag{4.41}$$

The results of such a clarifying calculation allow us to present the calendar time forecast (T_{CTF}) of the earthquake onset in the reference stations location as the next forecast range:

$$T_{CTF} \Rightarrow T_{LB} < T_{MPV} < T_{UB}. \tag{4.42}$$

4.7.4.2 Earthquake Strength Forecasting

The earthquake strength, measured in magnitudes M, is calculated, using the following formula [56]:

$$M = \log A - \log A_0. \tag{4.43}$$

According to [63], the A_0 level is assumed to be equal to 1 microns at a distance of 100 km from the standard earthquake epicentre.

We write the expression (4.42) for two earthquakes, assigning them conditional numbers 1 and 2, assuming the zero level A_0 is unchanged, both in magnitude and according to its registration conditions (seismograph type and remoteness from the epicentre).

$$M_1 = \log A_1 - \log A_0. \tag{4.44}$$

$$M_2 = \log A_2 - \log A_0. \tag{4.45}$$

Subtract Eq. (4.44) from Eq. (4.45).

$$M_2 - M_1 = \log A_2 - \log A_1. \tag{4.46}$$

We write the expression (4.46) in a more general form, removing index 2 for this purpose and taking the first earthquake as a standard one, replacing index 1 with the letter "S", respectively.

$$M = M_S + \log \frac{A}{A_S}. \tag{4.47}$$

The expression (4.47) allows you to determine the magnitude M of this earthquake, comparing it with the magnitude of the M_S earthquake, adopted as a standard.

Substituting in (11) the expression for the seismic wave tidal harmonic amplitude A_{SW} and assuming that $A = A_{SW}$ and $A_S = A_{SW}^S$, get a working formula to forecast the strength of the next earthquake.

$$M = M_S + \log \left[\frac{1}{N} \sum_{i=1}^{N} \left(\frac{A_{MTi}}{L_i} \right)^2 \right] - \log \left[\frac{1}{N_S} \sum_{k=1}^{N_S} \left(\frac{A_{MTk}^S}{L_{Sk}} \right)^2 \right]. \tag{4.48}$$

In the general case, the number of stations may not coincide ($N \neq N_S$) therefore, in (4.48) the total amplitude value is normalized by the number of reference stations, where measurements were taken during the preparation of the standard and next earthquake.

4.7.5 Results

The research results include the verification results of the above-described method and its approbation results when forecasting a ripening earthquake.

Figure 4.42 shows the reference seismic stations layout in three zones. In two of these zones (on the island of Hokkaido and the outskirts of Fukushima), destructive earthquakes occurred in 2003 and 2011. The measurements results performed in the territories were used in the method verification under consideration.

The considered forecasting method approbation was carried out according to the results of seismic signal measurements performed in the third zone, located in the area of the Nankai Trench (See Fig. 4.42).

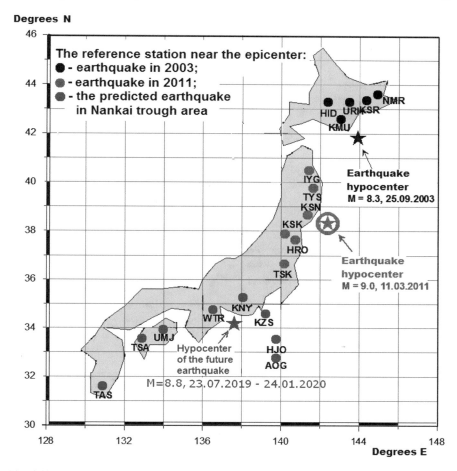

Fig. 4.42 Position and code numbers of Japanese seismic stations of the F-net network, which were considered as reference

4.7.5.1   The Verification Results of Forecasting Methods

Verification, as a procedure for assessing the forecast reliability, was carried out by retrospective parameters forecasting of previously occurring earthquakes and then comparing the forecasted and actually occurring parameters of earthquakes.

As the actual parameters were considered the parameters of two Japanese earthquakes that occurred in Hokkaido region on 25 September 2003 (the epicentre coordinates are 42.0° N, 142.0° E) and in the Fukushima area on 11 March 2011 (the epicentre coordinates are 38.32° N, 142.38° E).

Fig. 4.43 Approximation of the total tidal harmonic amplitude trend $A_{SV}(t_j)$ by graph of forecast model $A_{MT}^C(t_j)$. 1—total trend of the amplitude of the tidal harmonic $A_{SV}(t_j)$;2—graph of the forecasting model $A_{MT}^C(t_j)$

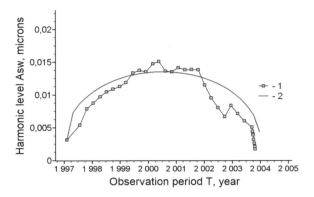

Fig. 4.44 Change in the forecast range during the observed period from 01/01/1997 to 09/25/2002. 1—upper bound of the forecast T_{UB}; 2—actual time of earthquake occurrence $T_{ACT} = 2002.83$ (25 September 2003); 3—most probable forecast T_{MPF} (standard deviation σ_F of the forecast is 0.067 years); 4—lower bound of the forecast T_{LB}

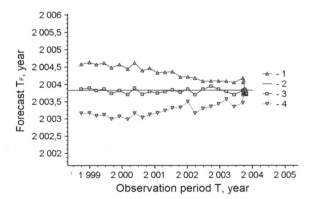

Earthquake Forecast Verification Near Hokkaido

During verification, the measurements results performed over a period of 6 years (from 1 January 1997 to 25 September 2003) were examined at five reference seismic stations located on the coast of Hokkaido Island: KMU, HID, KSR, URH and NMR (See Fig. 4.42).

Earthquake Date Forecast Verification

Earthquake forecast results, are shown in Figs. 4.43, 4.44, 4.45, 4.46.
    Forecasting results describe:

- change during the observation period from 01/01/1997 to 09/25/2003, the total trend of the tidal harmonic amplitude $A_{SW}(t_j)$ (1) and forecast model graph $A_{MT}^C(t_j)$ (2) (See Fig. 4.43);
- the change nature during the observation period of the forecast range and the most likely forecast value made without differentiating it into the accepted forecast types: long term, medium term and short term (See Fig. 4.44).

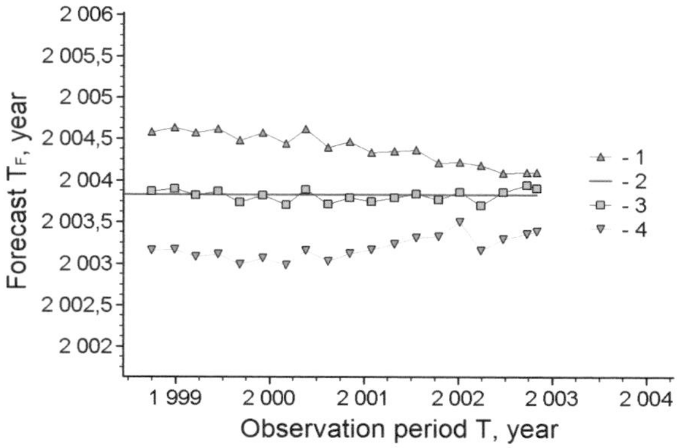

Fig. 4.45 Change in the forecast range during the long-term forecasting period from 01/01/1997 to 09/25/2002. 1—upper bound of the forecast T_{UL}; 2—actual time of earthquake occurrence $T_{ACT} = 2002.83$ (25 September 2003); 3—most probable forecast T_{MPF} (standard deviation σ_F of the forecast is 0.07 years); 4—lower bound of the forecast of T_{LB}

Fig. 4.46 Change in the forecast range during the medium-term forecasting from 09/25/2002 to 08/25/2002. 1—upper bound of the forecast T_{UL}; 2—actual time of earthquake occurrence $T_{ACT} = 2002.83$ (25 September 2003); 3—most probable forecast T_{MPF} (standard deviation σ_F of the forecast is 0.68 months); 4—lower bound of the forecast T_{LB}

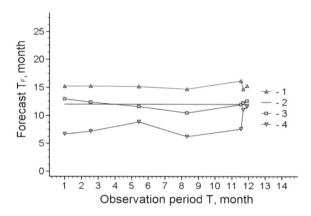

forecast results made in accordance with the traditional forecast types differentiation for the long term (See Fig. 4.45), medium term (See Fig. 4.46) and short term (See Fig. 4.47).

Protocol.
long-term forecasting.
from **2/16/1999.**
Forecasted Earthquake Date:
2002.887 year.
(10/31/2003 (16 h 56 min)).
changes in the following range:
2002.209…2003.566 years.
(from 2/15/2003 (9 h 34 min) to 6/24/2004 (23 h 53 min)).

The long-term forecasting protocol from **2/16/1999** shows that the most likely forecast T_{MPV} differs from the actual earthquake date T_{ACT} up for six days ($T_{MPV} = 10/31/2003 > T_{ACT} = 09/25/2003$).

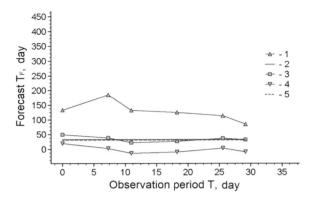

Fig. 4.47 Change in the forecast range during the short-term forecasting period from 25 August 2003 to 25 September 2002. 1—upper bound of the forecast T_{UL}; 2—actual time of earthquake occurrence $T_{ACT} = 2002.83$ (25 September 2003); 3—most probable forecast T_{MPF} (standard deviation σ_F of the forecast is 6.6 days); 4—lower bound of the forecast T_{LB}; 5—average of the most probable forecast $\overline{T}_{MPF} = 23$ September 2002

Long-term forecast verification shows that the rms value σ_F of the oscillations of the long-term forecast most probable value T_{MPV} calculated by the formula (4.41) varies within 0.067–0.07 years.

Medium-term forecast verification shows that the rms value σ_F of the oscillations of the medium-term most probable value T_{MPV} calculated by the formula (4.41) is 0.68 months.

Protocol.
medium-term forecasting.
from **8/31/2002.**
Forecasted Earthquake Date:
2002.772 year.
(09/08/2003 (22 h 21 min)).
changes in the following range:
2002.374…2003.168 years.
(from 04/15/2003 (7 h 8 min) to 02/01/2004 (14 h 58 min)).

From the protocol for medium-term forecasting, performed almost a year before the actual earthquake date, it follows that the most likely forecast T_{MPV} differs from the actual date in the lower side by 17 days ($T_{MPV} = 09/08/2003 < T_{ACT} = 09/25/2003$).

Protocol.
short-term forecasting.
from **09/14/2002.**
Forecasted Earthquake Date:
2002.801 year.
(09/19/2003 (18 h 11 min)).

Fig. 4.48 Change during the observation period from 01/01/1997 to 09/25/2003 in the forecast of the northern latitude of the epicentre of the earthquake that occurred on 25 September 2002. 1—coordinate forecast N_{ACT} (the rms value σ_N of the latitude forecasting is 0.16°); 2—actual value of the coordinate $N_{ACT} = 42°$

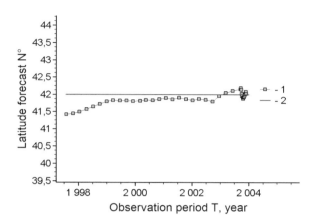

Fig. 4.49 Change during the observation period from 01/01/1997 to 09/25/2003 in the forecast of the eastern longitude of the epicentre of the earthquake that occurred on 25 September 2002. 1—coordinate forecast E_F (the standard deviation σ_E of the longitude forecast is 0.27°); 2—actual value of the coordinate $E_{ACT} = 143°$

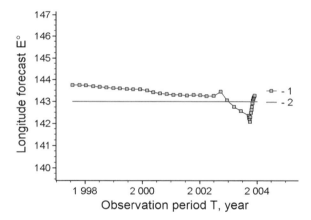

changes in the following range:
2002.796…2002.807 year.
(from 09/17/2003 (16 h 7 min) to 09/21/2003 (19 h 56 min)).

Short-term forecast verification shows that the rms value σ_F of the oscillations of the short-term most probable value T_{MPV} calculated by the formula (6) is 6.6 days.

From the short-term forecasting protocol, performed 11 days before the actual earthquake date, it follows that the most likely forecast T_{MPV} differs from the actual T_{ACT} date downwards by 6 days ($T_{MPV} = 09/19/2003 < T_{ACT} = 09/25/2003$).

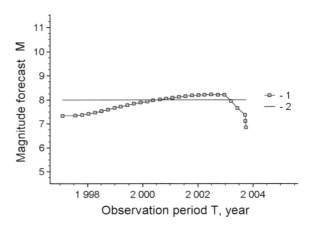

Fig. 4.50 Change during the observation period from 01/01/1997 to 09/25/2003, the forecast of the magnitude of the earthquake that occurred on 25 September 2003 1—forecast of the earthquake strength M_F (standard deviation σ_M of the earthquake magnitude is 0.32); 2—actual magnitude $M_{ACT} = 8$

Forecast Verification of the Earthquake Epicentre Coordinates

The forecast (See Figs. 4.48 and 4.49) shows that the deviation of the epicentre coordinates forecast N_F and E_F from their actual values N_{ACT} and E_{ACT} is characterized by the rms value σ varying from 0.16° to 0.27°, which is in terms of distance, respectively, 18 and 30 km.

These distances correspond to the accuracy in determining the earthquake epicentres coordinates acceptable in seismology [63].

Forecast Verification of the Earthquake Strength

Figure 4.50 shows that the earthquake magnitude forecast M_F, calculated according to the formula (13), fluctuates relative to the actual magnitude M_{ACT}. An earthquake that occurred in Japan in 2011 was considered as a standard. The rms σ_M of the forecast fluctuations is 0.32 magnitude and is 4% of the earthquake strength actual characteristic ($M_{ACT} = 8$).

Earthquake Forecast Verification in the Area of Fukushima

During verification, the measurements results performed over a period of 10 years (from 1 January 2004 to 11 March 2011) were examined at six reference seismic stations located in the area of Fukushima: TSK, HRO, KSK, KSN, TYS and IYG (See Fig. 4.42).

Earthquake Date Forecast Verification

Earthquake forecast results are shown in Figs. 4.51, 4.52, 4.53, 4.54, 4.55 and describe.

Fig. 4.51 Approximation of the total tidal harmonic amplitude trend $A_{SW}(t_j)$ by graph of forecast model $A^C_{MT}(t_j)$. 1—total trend of the amplitude of the tidal harmonic $A_{SW}(t_j)$; 2—graph of the forecast model $A^C_{MT}(t_j)$

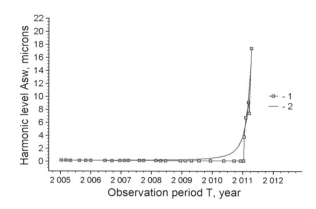

Fig. 4.52 Change in the forecast range during the observed period from 1.01.2004 to 11.02.2011. 1—upper bound of the forecast T_{UL}; 2—actual time of earthquake occurrence $T_{ACT} = 2011.28$ (11 March 2011); 3—most probable forecast T_{MPF} (mean square deviation of the forecast is 0.0118 years; 4—lower bound of the forecast of T_{LB}

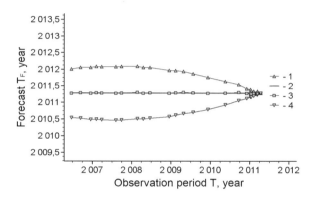

Fig. 4.53 Change in the forecast range during long-term forecasting in the period from 01/01/2004 to 03/11/2010. 1—upper bound of the forecast T_{UL}; 2—actual time of earthquake occurrence $T_{ACT} = 2011.28$ (11 March 2011); 3—most probable forecast T_{MPF} (standard deviation σ_F of the forecast is 0.01 years); 4—lower bound of the forecast T_{LL}

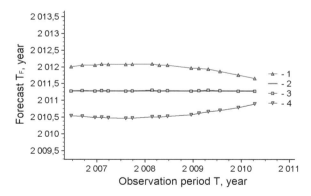

the change during the observation period from 1 January 2004 to 11 March 2011 of the tidal harmonic amplitude trend $A_{SW}(t_j)$ and approximate graph of the forecast model $A^C_{MT}(t_j)$ (See Fig. 4.51);

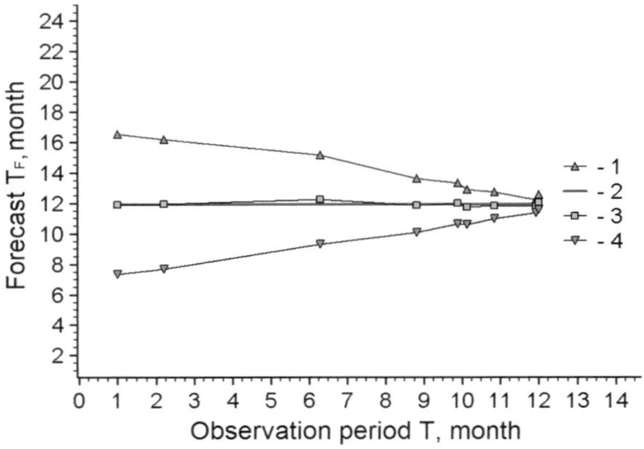

Fig. 4.54 Change in the
forecast range during the
medium-term forecasting in
the period from 03/11/2010
to 02/11/2011. 1—upper
bound of the forecast T_{UL};
2—actual time of earthquake
occurrence $T_{ACT} = 2011.28$
(11 March 2011); 3—most
probable forecast T_{MPF}
(mean square deviation σ_F
of the forecast is
0.30 months); 4—lower
bound of the forecast T_{LB}

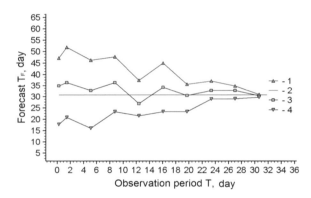

Fig. 4.55 Change in the
forecast range during
short-term forecasting in the
period from 02/11/2011 to
03/11/2011. 1—upper bound
of the forecast T_{UL};
2—actual time of earthquake
occurrence $T_{ACT} = 2011.28$
(11 March 2011); 3—most
probable forecast T_{MPF}
(mean square deviation σ_F
of the forecast is 2.9 days);
4—lower bound of the
forecast T_{LB}

the change nature during the observation period of the forecast range and the most
likely forecast value made without differentiating it into the accepted forecast
types: long term, medium term and short term (See Fig. 4.52);

forecast results made in accordance with the traditional forecast types differentiation for the long-term (See Fig. 4.53), medium-term (See Fig. 4.54) and short-term
(see Fig. 4.55).

Protocol.
long-term forecasting.
from **11/14/2006**.
Forecasted Earthquake Date:
2011.252 year.
(03/01/2011 (18 h 11 min)).
changes in the following range:
2010.471…2012.034 years.
(from 5/20/2010 (16 h 17 min) to 01/12/2012 (7 h 2 min)).

Long-term forecast verification shows that the rms value σ_F most probable forecast value T_{MPV} calculated by the formula (6) varies within 0.010–0.018 years.

The long-term forecasting protocol from 11/14/2006 shows that the most likely forecast T_{MPV}, made more than 4 years before the earthquake, differs from the actual earthquake date T_{ACT} down by 8 days ($T_{MPV} = 03/01/2011 < T_{ACT} = 03/11/2011$).

Medium-term forecast verification shows that the rms value σ_F of the oscillations of the medium-term most probable value T_{MPV} is 0.30 months.

Protocol.
medium-term forecasting.
from **11/09/2009**.
Forecasted Earthquake Date:
2011.293 year.
(03/16/2011 (12 h 8 min)).
changes in the following range:
2010.800…2011.785 years.
(from 09/18/2010 (10 h 38 min) to 09/13/2011 (15 h 31 min)).

From the protocol for medium-term forecasting from **11/09/2009**, performed almost a year before the actual earthquake date T_{ACT}, it follows that the most likely forecast T_{MPV} differs from the actual date in a big way by 5 days ($T_{MPV} = 03/16/2011 > T_{ACT} = 03/11/2011$).

Protocol.
short-term forecasting.
from **03/09/2011**.
Forecasted Earthquake Date:
2011.281 year.
(03/11/2011 (12 h 53 min)).
changes in the following range:
2011.278…2011.284 year.
(from 03/10/2011 (9 h 9 min) to 03/13/2011 (14 h 13 min)).

Short-term forecast verification shows that the rms value σ_F of the oscillations of the short-term most probable value T_{MPV} is 2.9 days.

From the short-term forecasting protocol from **03/09/2011**, performed 2 days before the actual earthquake date T_{ACT}, it follows that the most likely forecast T_{MPV} coincides with the actual earthquake date ($T_{MPV} = 03/11/2011 = T_{ACT} = 03/11/2011$).

Forecast Verification of the Earthquake Epicentre Coordinates

The calculation results show (See Figs. 4.56 and 4.57) that the epicentre coordinates forecast N_F и E_F varies around their actual values N_{ACT} и E_{ACT}. The rms value σ of the coordinates forecast varies within $0.54° - 0.99°$. As a percentage relative to the actual coordinate's values, this is, respectively, 0.38%–2.6%.

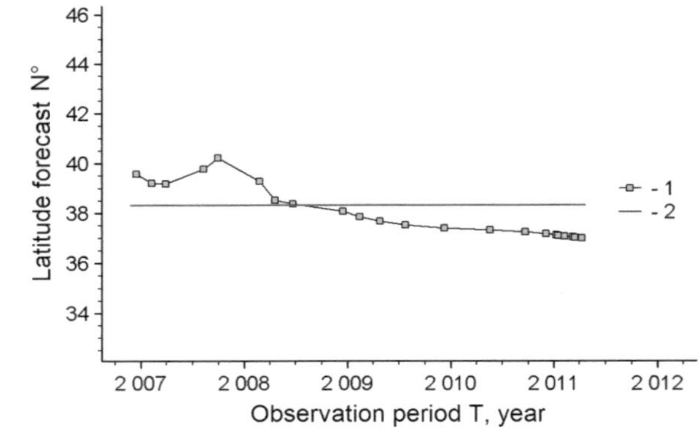

Fig. 4.56 Change during the
observation period from
01/01/2004 to 03/11/2011 of
the forecast of northern
latitude N° epicentre of the
earthquake that occurred on
11 March 2011.
1—coordinate forecast N_F
(standard deviation σ_N of the
latitude forecast is 0.99°);
2—actual value of the
coordinate $N_{ACT} = 38.32°$

Fig. 4.57 Change during the
observation period from
01/01/2004 to 03/11/2011 of
the forecast of the eastern
longitude E of the epicentre
of the earthquake that
occurred on 11 March 2011.
1—coordinate forecast E_F
(standard deviation σ_E of the
latitude forecast verification
of the earthquake strength
ion is 0.54°); 2—the actual
value of the coordinate E_{ACT}
$= 142.37°$

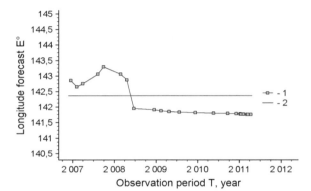

Forecast Verification of the Earthquake Strength

Figure 4.58 shows that the earthquake magnitude forecast M_F, calculated according to the formula (13), fluctuates relative to the actual magnitude M_{ACT}. An earthquake that occurred in Japan in 2003 was considered as a standard. The rms σ_M of the forecast fluctuations is 0.29 magnitude and is 2.2% of the earthquake strength actual characteristic ($M_{ACT} = 9$).

Thus, the verification results testify to the stability and insignificant variability throughout the entire period of seismic signals observation of the time forecasts, place and earthquake strength. Moreover, the forecasts each of the three earthquake parameters can be obtained long before the occurrence of the forecasted natural disaster; as a result, there is no need to apply the long-term, medium-term and short-term forecasting standard accepted in seismology.

Successful verification results of the earthquake forecasting method proposed by the author served as the basis for testing the methodology by forecasting the parameters of a future earthquake.

Fig. 4.58 Change during the observation period from 01/01/2004 to 03/11/2011 for the forecast of the magnitude of the earthquake that occurred 11 March 2011. 1—forecast of the earthquake force of the M_F (the mean square deviation σ_M of the magnitude forecast is 0.29, 2—is the actual magnitude value $M_{ACT} = 9$

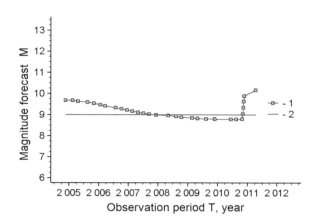

Fig. 4.59 Approximation of the total tidal harmonic amplitude trend $A_{SV}(t_j)$ by graph of forecast model $A^C_{MT}(t_j)$. 1—total trend of the amplitude of the tidal harmonic $A_{SV}(t_j)$; 2—graph of the forecast model $A^C_{MT}(t_j)$

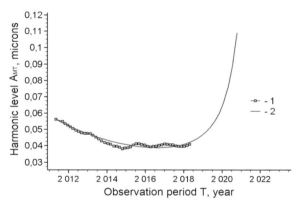

Fig. 4.60 Change in the forecast range during the observed period from 01/01/2011 to 03/25/2018. 1—upper bound of the forecast T_{UB}; 2—most probable forecast T_{MPF} (mean square deviation σ_F of the forecast is 0.9 years); 3—averaged value of the most probable forecast T_{MPF} = 2019.59 (2 July 2019); 4 – lower bound of the forecast T_{LB}

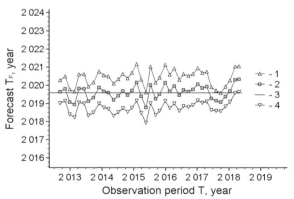

4.7.6 The Approbation Results of the Forecasting Method

During the approbation of the forecasting method, the measurements results performed at seven reference seismic stations located in the Nankai Trench area were considered: TAS, AOG, HJO, UMJ, KZS, WTR and KNY (See Fig. 4.42). Measurements were taken from 01/01/2012 to 03/25/2018.

4.7.6.1 Earthquake Forecasting Time

Protocol.
  long-term forecasting.
  from **3/27/2018**.
  Forecasted Earthquake Date:
  2019.646 year.
  (07/23/2019 (18 h 43 min)).
  changes in the following range:
  2019.142…2020.148 years
  (from 01/22/2019 (22 h 28 min) to 01/24/2020 (13 h 34 min)).

The long-term forecast approbation shows that the rms value σ_F most probable forecast value T_{MPV} calculated by the formula (6) is 0.16 year.

The long-term forecasting protocol from **27.2.2018** shows that the most likely forecast for the earthquake date is 07/23/2019 and varies from 01/22/2019 to 01/24/2020.

Fig. 4.61 Change during the observation period from 01/01/2011 to 03/25/2018, the forecast of the northern latitude of the epicentre of the ripening earthquake. 1—coordinate forecast $N°_F$ (standard deviation σ_N of the latitude forecast is 0.89°); 2—averaged value of the coordinate $N_{AVE} = 33.34°$

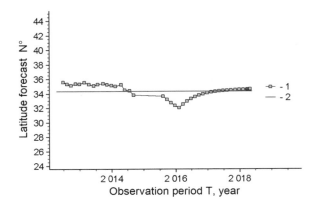

Fig. 4.62 Change during the observation period from 01/01/2011 to 03/25/2018, the forecast of the eastern longitude of the epicentre of the ripening earthquake. 1—coordinate forecast $E^\circ{}_F$ (standard deviation σ_E of the longitude forecast is 2.53°); 2—averaged value of the coordinate $E_{AVE} = 137.55°$

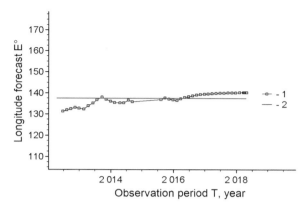

4.7.6.2 Forecasting of the Earthquake Epicentre Coordinates

The calculation results (See Figs. 4.61 and 4.62) show that the rms value σ of the coordinates forecast varies within $0.89° - 2.53°$. As a percentage relative to their average value, this is, respectively, 2.6%–1.8%.

The averaged values of the forecast earthquake epicentre coordinates are 33.34 degrees north latitude and 137.55 degrees east longitude.

4.7.6.3 Forecasting the Strength of a Ripening Earthquake

Figure 4.63 shows that the average magnitude forecast of earthquake M_F, calculated according to the formula (13), is equal to 8.84. An earthquake that occurred in Japan in 2011 was considered as a standard. The rms σ_M of forecast fluctuations relative to this value is 0.035 magnitude, which is 0.4% of the magnitude average value $M_{AVE.}$

Fig. 4.63 Change during the observation period from 01/01/2011 to 03/25/2018, the forecast of the power of the ripening earthquake. 1—forecast of the earthquake force of the M_F (standard deviation σ_M of the magnitude forecast is 0.035); 2—averaged value of the magnitude forecast $M_{AVE} = 8.84$

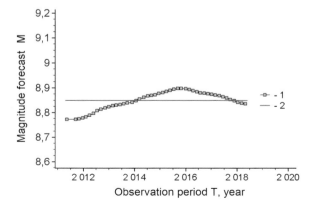

Fig. 4.64 Change during the
period of observation of the
magnitude forecast:
1—earthquake, held on 28
September 2003;
2—earthquake, held on 11
March 2011; 3—ripening
earthquake

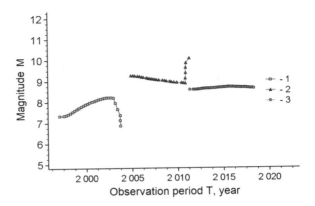

In Fig. 4.64 for comparison, retrospective forecasts of earthquake magnitudes, obtained during forecasting method verification, and forecast of the ripening earthquake strength, expected in the period from 01/22/2019 to 01/24/2020, are given. The comparison shows that the next earthquake is intermediate in relation to the occurred earthquakes.

Forecast ranges and the change in the current date are matched in Fig. 72 (indicated by the dotted straight line in the Fig.). As you can see, at the forecast time between the most probable date of the ripening earthquake (23 July 2019) and the forecast date (25 March 2018), there is more than a year reserve.

During this period, continuous monitoring should be carried out with the aim of improving the forecast accuracy of the seismic signal level at each of the reference seismic stations.

Figure 4.66 shows two trends of the tidal harmonics amplitude $A_{SW}(t_j)$. The first was recorded during the earthquake preparation, which was in Japan on 11 March 2011 and is used as a calibration graph for the second trend currently observed. For the convenience of comparing trends, the time is plotted on the abscissa axis,

Fig. 4.65 Comparison of the
forecast of the earthquake
time and the current calendar
time: 1—earthquake, held on
28 September 2003;
2—earthquake, held on 11
March 2011; 3—ripening
earthquake

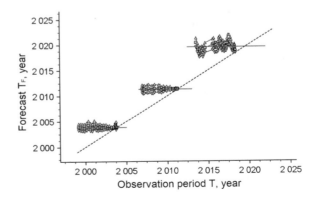

Fig. 4.66 Comparison of the trends of the amplitude of the tidal harmonic of A_{SV}. 1—amplitude change in the period from 01/01/2004 to 03/11/2011 (calibration trend); 2—amplitude change in the period from 01/01/2011 to 03/25/2018 (trend, observed in the current period of time)

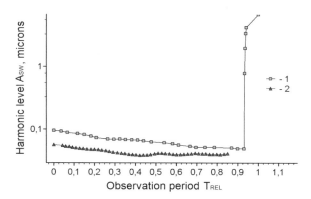

normalized by the earthquakes onset time (held on 11 March 2011 and ripening—27 July 2019).

The tidal harmonic trend of the ripening earthquake observed in the current time period is visually, practically, equidistant to the stationary part of the calibration trend. Equidistance is quantitatively confirmed by the high level of compared trends correlation ($R = 0.98976$, See Fig. 4.67).

$$(A_{SW19} = 0.01967 + 0.39298 \cdot A_{SW11}, R = 4.0.98976)$$

The research results conducted by the author in the methods development for forecasting earthquakes showed that in trend behaviour the nature of the earthquake prognostic feature is an effective and very informative forerunner of an earthquake. What, in fact, was confirmed by the verification results and approbation of earthquake forecasting methods given in the article, which in this case is based on the seismic oscillations analysis of the earth's surface. Earthquake forecasting method is protected by three national patents [64–66].

Fig. 4.67 Regression relationship between the calibration trend and the trend of the ripening earthquake recorded, respectively, in the periods from 01/01/2004 to 03/11/2011 (A_{SW11}) and from 01/01/2011 on 25 March 2013 (A_{SW19}). 1—experimental data; 2—is a graph of correlation dependence

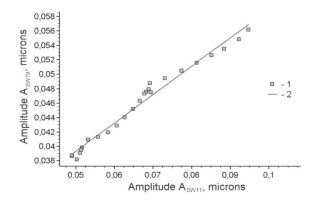

The trend behaviour, as a prognostic feature, fully meets the requirements for them, given in the introduction, namely.

- The trend has a clear physical meaning, naturally describing the preparation period of the next earthquake, considered in the article the behaviour of prognostic feature—tidal harmonic amplitudes.
- The trend relationship, as a prognostic feature, with the earthquakes preparation process is physically grounded.
- As a prognostic feature, the trend is provided with observational data, both in time, due to the presence of long-term values series, for example, tidal harmonic amplitudes, and in space due to the use of several reference seismic stations located in a seismic zone, which ensures the monitoring results mapping of the country's seismic situation.
- The presence of a trend model describing its behaviour, both during the earthquake preparation and at the time of its anomalies, formalizes the procedure for forecasting the moment, epicentre coordinates and the earthquake strength.
- The trend allows, if necessary, to obtain traditional estimates of the retrospective statistical characteristics of each precursor, whose behaviour it describes: probabilities of successful forecasts (detection probabilities), probabilities of false alarms, forecast efficiency (informativeness), etc.

To take full benefits advantage of this earthquake precursor, you need to consistently perform the following:

- as a continuous monitoring result of the indications of the seismic stations national network, identify areas with an increased seismic signal relative to the surrounding territory and consider the seismic stations located in this zone as reference;
- carry out signal recorded spectral analysis by each of the reference stations, highlight the tidal harmonic in the spectrum's ultra-low-frequency part and determine its amplitude, simultaneously fixing the current measurement date;
- compile for each of the reference seismic stations on the basis of parameters pair "tidal harmonic amplitude—date of seismic signal recording" the trend of the tidal harmonic amplitude;
- solve the equations system (3), determining from the results of this and complementary calculations using formulas (6) and (12), the desired parameters of a future earthquake: T_F, x_O, y_O, M.

References

1. ISO 10816–1: Mechanical vibration. Evaluation of machine vibration by measurements on non-rotating parts (1995)
2. Chetyrkin, E.M.: Statistical Methods of Forecasting, p. 200. Statistics, Moscow (1977)
3. Ivahnenko, A.G., Yurachkovskiy, Yu.P.: Complex Systems Simulation by Experimental Data, p. 120. Radio i svyaz, Moscow (1987)
4. Pronikov, A.S.: Reliability Machines, p. 592. Mechanical Engineering, Moscow (1978)

5. Lvivskiy, E.N.: Statistical Methods for Constructing Empirical Formulas, p. 239. Vysshaja shkola, Moscow (1988)
6. Greshilov, A.A., Stakun, V.A., Stakun, A.A.: Mathematical Methods of Forecasting, p. 112. Radio i svyaz, Moscow (1997)
7. Kudritskiy, V.D.: Forecasting the Reliability of Radio-Electronic Devices, p. 156. Kiev, Technique (1973)
8. Davydov, P.S.: Technical Diagnostics of Radio-Electronic Devices and Systems, p. 256. Radio i svyaz, Moscow (1988)
9. Zhernakov, S.V.: Trend analysis of parameters of aviation gas turbine engine based on neural network technology. Bull. UGATU Thermal Electro-rocket Eng. Power Plan TCR LA Ufa, Russia 14(4), 25–32 (2011). ISSN 1992-6502.
10. Davydov, P.S., Ivanov, P.A.: Operation of Aviation Radio-Electronic Equipment, p. 240. Transport, Moscow, Reference book (1990)
11. Popkov, V.I., Myshinsky, E.L., Popkov, O.I.: Vibroacoustic Diagnostics in Shipbuilding, p. 256. St. Petersburg, Shipbuilding (1982)
12. Shchepin, L.S., Zaripov, R.M.: The way to increase the reliability of the centrifugal pumping unit of hydrocarbon raw materials and the system for diagnosing iTCR technical condition. Russian Federation Patent 2068553, 27 June 2009 (2009)
13. Cantilever pumps. https://www.agrovodcom.ru/konsol_pump.php Accessed 1 Jan 2003.
14. Shannon, K.: Works on Information Theory and Cybernetics, p. 830. Foreign Literature Publisher, Moscow (1963)
15. Sedov, L.I.: Methods of Similarity and Dimension in Mechanics, p. 440. Nauka, Moscow (1977)
16. Zadeh, L.A.: The Concept of a Linguistic Variable and iTCR Application to the Adoption of Approximate Solutions, p. 165. Mir, Moscow (1976)
17. An act of technical investigation of the causes of the accident that occurred on August 17, 2009 in the branch of the Open Joint-Stock Company "RusHydro"—Sayano-Shushenskaya HPP named after PS Neporozhny
18. STO 70238423.27.140.001–2011: Hydroelectric power stations. Techniques for assessing the technical condition of the main equipment
19. Vasiliev, Y.S., Petrenya, Y.K.: Georgievskaya 22(2), (2017). ISSN 184–203, 2542–1239
20. Sanko, A.A., Tyupin, R.L., Sheinikov, A.A.: The effectiveness of the methods of trend control and forecasting of the technical condition of the gas turbine engine of a helicopter according to vibration measurements (2014). https://www.rusnauka.com/3_ANR_2014/Tecnic/9_156440.doc.htm
21. Lopatin, A.S.: Justification of diagnostic signs of rotor imbalance. Nat. Acad. Sci. Ukraine Institute Electric Weld. 2 (2002). ISSN 36-39, 0235-3474
22. Barkov, A.V.: Monitoring and Diagnostics of Rotary Machines by Vibration, p. 159. St. Petersburg State Marine Technical University, St. Petersburg (2000)
23. Palmgren, A.G.: Die Lebensdauer von Kugellagern, Zeitschrift des Vereines Deutscher Ingenieure (ZVDI), Berlin, Germany 68(14), 339–341 (1924)
24. Miner, M.A.: Cumulative damage in fatigue. J. Appl. Mech. USA 3 (1945). ISSN 159-164, 0021-8936
25. Dimla, , D.E., Lister, P.M., Leighton, N.J.: Neural network solutions to the tool condition monitoring problem in metal cutting—a critical review of methods. Int. J. Mach. Tools Manuf. Amsterdam, Netherlands 37 (1997). ISSN 1219-1241, 0890-6955
26. Byrne, G., Dornfeld, D., Inasaki, I., Ketteler, G., König, W.: Teti 44 (1995). ISSN 541-567, 0007-8506
27. Sirkant, R., Subrahmanyam, S., Chen, K. Krishna, V.P.: Experimental selection of special geometry cutting tool for minimal tool wear, Maribor, Slovenia, vol. 5, 2010, p. 13–23, ISSN 1854–6250.
28. Castejon, M., Alegre, E., Barreiro, J., Hernandez, L.K.: On-line tool wear monitoring using geometric descriptors from digital images. Int. J. Mach. Tools Manuf. Amsterdam, Netherlands 47 (2007). ISSN 1847-1852, 0890-6955

29. Ghani, J.A., Rizal, M., Nuawi, M.Z., Ghazali, M.J., Haron, C.H.C.: Monitoring online cutting tool wear using low-cost technique and user-friendly GUI. Wear, Amsterdam, Netherlands **271** (2011). ISSN 2619-2623, 0043-1648
30. ISO 3685:1003 (E): Tool-life testing with single-point turning tools (1992)
31. Scheffer, C., Heyns, P., S.: An industrial tool wear monitoring system for interrupted turning. Mech. Syst. Signal Process. Amsterdam, Netherlands **18** (2004). ISSN 1219–1242, 0888–3270
32. Shao, H., Wang, H.L., Zhao, X.M.: A cutting power model for tool wear monitoring in milling. Int. J. Mach. Tools Manuf. Amsterdam, Netherlands **44** (2004). ISSN 105-108, 0890-6955
33. Aruvali, T., Serg, R., Otto, T.: In-process vibration monitoring on CNC lathe. In: 10th International Symposium Topical Problems in the Field of Electrical and Power Engineering, pp. 174–178. Estonia, Pärnu (2011)
34. Haber, R.E.: Alique **13** (2005). ISSN 825-849, 0957-4158
35. Ostasevicius, V., Gaidys, R., Rimkeviciene, J.: Dauksevicius **329** (2010). ISSN 4866–4879, 0022-460X
36. Chen, Ch., Wang, Y.C., Lee, B., Y.: The effect of surface roughness of end mills on optimal cutting performance for high-speed machining **59** (2013). ISSN 124-133, 0039-2480
37. Velayudham, A., Krishnamurthy, R., Soundarapandian **412** (2005). ISSN 141-144, 0921-5093 (2005)
38. Antić, A., Kozak, D., Kosec, B., Šimunović, G., Šarić, T., Kovačević, D.: Čep **20** (2013). ISSN 105-112, 1330-3651
39. Narasimha, M., Sridhar, K., Kumar, R.R., Kassie, A.A., Improving cutting tool life a review. Int. J. Eng. Res. Develop. **7** (2013). ISSN 67-74, 2278-2800
40. Nikhare, P., Ragai, I., Loker, D., Sweeney, S., Conklin, C., Roth, J.T.: Investigation of acoustic signals during W1 tool steel quenching. In: Proceedings of the ASME International Manufacturing Science and Engineering Conference, pp. 1–9. Charlotte, NC, USA (2015). ISBN 978–0–7918–5683–3.
41. Erich, N.J., Nikhare, C.P., Conklin, C., Loker, D.R.: Loker, Study of acoustic signals and mechanical properties dependence during cold drawn A36 steel quenching. In: Proceedings of the International Deep Drawing Research Group Conference, pp. 338–346. Shanghai, China (2015). ISBN 978–1–5108–4896–2.
42. Jaeckel, O.: Strengths and weaknesses of calculating beam forming in the time domain. In: Proceedings of the 1st Berlin Beam Forming Conference, pp. 1–10. Berlin, Germany (2006)
43. Romenskiy, I., Jaeckel, O.: Improvement of source separation for phased microphone array measuremenTS. In: Proceedings of the 2nd Berlin Beam Forming Conference, pp. 1–8. Berlin, Germany (2008)
44. Twardowski, P., Wojciechowski, S., Wieczorowski, M., Mathia, T.: Surface roughness analysis of hardened steel after high-speed milling, Scanning, London, United Kingdom **33** (2011). ISSN 386–394, 0161–0457
45. Sapozhkov, M.A.: Electroacoustics Communication, Moscow, 1978, p. 275
46. Ananiev, V.A.: A System of Ventilation and Air Conditioning, p. 416. Euroclimate, Moscow (2001)
47. Nahornyi, V., Aleksenko, O., Fedotova, N.: Adaptive control of the metalworking technology systems operation based on the forecast of the actual resource of the cutting tool. In: 22nd International Conference Information and Software Technologies, pp. 187–198. ICIST, Druskininkai, Lithuania (2016)
48. Guzij, A.G., Bogomolov, A.V., Kukushkin, J.A.: Theoretical foundations of the functional-adaptive control of man-machine systems of increased accident rate. Mechatron. Autom. Control **1**, 39 (2005). ISSN 1684-6427
49. Monmollen, M.: Man-machine Systems, p. 256. Mir, Moscow (1973)
50. Sheridan, T.B., Ferrell, U.R.: Man-Machine Systems: Information, Control and Decision Models of Human Performance, p. 440. MIT Press, Cambridge, Massachusetts (1974)
51. Asan: **58** (2017). ISSN 301-307, 0003-6870
52. Rothmore, P., Aylward, P., Oakman, J., Tappin, D., Gray, J., Karn, J.: The stage of change approach for implementing ergonomics advice. Appl. Ergon. Amsterdam, Netherlands **59** (2017)

53. Karltun, A., Karltun, J., Berglund, M.: Eklund **59** (2017). ISSN 182-190, 0003-6870
54. O'Neill, J., O'Neill, D.A., Lewinski, W.J.: Toward a taxonomy of the unintentional discharge of firearms in law enforcement. Appl. Ergon. Amsterdam, Netherlands **59** (2017). ISSN 283-292, 0003-6870
55. Regente, J., Zeeuw, J., Bes, F., Nowozin, C., Appelhoff, S., Wahnschaffe, A., Münch, M., Kunz, D.: Can short-wavelength depleted bright light during single simulated night shifts prevent circadian phase shifts? Appl. Ergon. Amsterdam, Netherlands **61**, 22–30 (2017). ISSN 0003-6870.
56. Albin, T.J.: Design with limited anthropometric data: A method of interpreting sums of percentiles in anthropometric design, Appl. Ergon. Amsterdam, Netherlands **62** (2017). ISSN 19-27, 0003-6870
57. Holtermann, A., Schellewald, V., Mathiassen, S.E.: A practical guidance for assessments of sedentary **63** (2017). ISSN 41-52, 0003-6870
58. Nahornyi, V. M.: Method of determining the residual service life of the machine, Ukraine Patent, UA 38438A, 15 May 2001 (2001)
59. Nahornyi, V. M.: Method of vibration diagnostics of the technical condition of the machine, Ukraine Patent, UA 51154A, 15 Nov 2002 (2002)
60. Nahornyi, V.M.: A device for automatic diagnostics of the hydraulic machines technical condition, USSR Patent, No.1763717, 23 Sep 1992 (1992)
61. Bestuzhev-Lada, I.V.: Forecasting Workbook, p. 430. Mysl', Moscow (1982)
62. Rikitake, T.: Forecasting of Earthquakes, p. 388. Mir, Moscow (1978)
63. Zavyalov, A.D.: Medium-Term Earthquake Forecast: Fundamentals, Methods, Implementation, p. 254. Nauka, Moscow (2006)
64. Lyubushin, A.A.: Prognostic properties of random fluctuations of geophysical characteristics. Interdiscip. Sci. Appl. J. Biosphere St. Petersburg, Russia **6**, 319–338 (2013). ISSN 2077-1371
65. Yurkov, Y.F., Gittis, V.G.: On the connection of seismicity with tidal wave phases 3(2005). ISSN 4–14, 0002–3337
66. Stepanov, N.N.: Spherical Trigonometry, p. 154. OGIZ, Moscow (1948)

Chapter 5
Conclusion

Forecasting the onset of the critical state of various events and phenomena, different in their nature and occurrence, is an important and not fully resolved issue of humanity.

The currently adopted method of forecasting mechanical systems resource, for example in engineering, which consists in extrapolating the graph of analytical dependence (forecast model) to the moment of intersection with the maximum permissible level of the information signal, is far from perfect.

This is due to the fact that information about the signal level is either absent or applies to a rather limited class of products for which similar data, for example, are available in the form of "Vibration Standards". However, even the existence of "norms …" does not guarantee from errors of forecasting the resource similar in design products for the reason that these norms were developed for objects that significantly differ from modern ones in specific load, structural materials, manufacturing technology, etc.

In contrast, the monograph considers a fundamentally different forecasting methodology allowing for the first time in world practice to solve the problem in each of the areas of human activity.

In the field of technology, the solution to the problem of forecasting the resource of modern mechanical systems was found in replacing information on the maximum permissible level of the information signal with information about the coefficients of the forecast model describing the trend of the information signal generated by a product during the entire observed (inter-repair) period.

For the first time, a resource forecast does not require knowledge of the average statistical data on the maximum permissible values of the controlled parameter. In accordance with the new forecasting methodology, the operating time between products to failure is determined by the results of the parametric identification of the trend model of the controlled parameter, one of the coefficients of which is the required resource of the supervised equipment. Testing of the proposed methodology for forecasting events and phenomena in nature and medicine proves its unique capabilities.

© The Author(s), under exclusive license to Springer Nature Switzerland AG 2021   95
A. Panda and V. Nahornyi, *Forecasting Catastrophic Events in Technology, Nature and Medicine*, SpringerBriefs in Computational Intelligence,
https://doi.org/10.1007/978-3-030-65328-6_5

The methodology allows for early diagnosis of a developing disease as well as to forecast the time of the occurrence of a devastating earthquake. The methodology is protected by ten patents and is implemented as a software product [1–6].

The software product consists of a central module, which implements the operation of parametric identification of a forecast model, and a series of auxiliary modules that collect initial data to analyse information about the behaviour of the control object and visualize the forecasted results on the device screen. Auxiliary modules are selected depending on the specifics of the problem to be solved (specificity and type of control object) [7–12].

The software product is designed for a variety of stationary and mobile computing devices ensuring the implementation of the proposed methodology into the wide practice of forecasting events and phenomena of various origins [13–18].

Prospects for further research are to create the universal control system for any technical, biological or natural system, the functioning of which is accompanied by the generation of control parameters of different physical nature [19–23].

Provided that these parameters objectively reflect the degree of criticality of the state of the control object and therefore can serve as a basis for forecasting the individual behaviour of the object until it reaches its actual state in the actually established conditions of its operation.

References

1. Nahornyi, V.V.: The long-term forecasting method of the coordinates of the next earthquake. Ukraine Patent, UA 124943, 25 (2018).
2. Nahornyi, V.V.: Method for forecasting the time of the next earthquake. Ukraine Patent, UA 126808, 10 (2018).
3. Nahornyi, V.V.: Method for forecasting the strength of the next earthquake. Ukraine Patent, UA 127519, 10 (2018).
4. Nahornyi, V.V.: Vibration diagnostics method of the metalworking machine technical state. Ukraine Patent, UA 91643, 10 (2014).
5. Nahornyi, V.M.: Method for determining the machine residual life. Ukraine Patent, UA 38438A, 15 (2001).
6. Panda, A., Nahornyi, V., Pandová, I., Harničárová, M., Kušnerová, M., Valíček, J., Kmec, J.: Development of the method for forecasting the resource of mechanical systems. Int. J. Adv. Manuf. Technol. Springer London Ltd., England **104** (2019). ISSN 0268–3768.
7. Panda, A., Nahornyi, V., Valíček, J., Harničárová, M., Pandová, I., Borzan, C., Cehelský, S., Androvič, L., Hakan, T., Kušnerová, M.: Application of cardio-forecasting for evaluation of human—operator performance. Int. J. Environ. Res. Public Health MDPI AG, Basel, Switzerland **17**(1), 12 (2020). ISSN 16604601.
8. Nahornyi, V.V.: Control of the Dynamic State of the Metal of the Technological System and Forecasting its Resource, p. 242. Sumy State University, Sumy (2016).
9. Nahornyi, V.V.: Forecasting of the moment of replacement of the cutting instrument by the sound level generated by the cutting process. In: Materials of the 1st Scientific and Practical Conference Innovations, Quality and Service in Engineering and Technology, Kursk, Russia, pp. 107–111 (2012). ISBN 978–5–9906896–1–9.
10. Nahornyi, V. V., Zaloga, V. A.: Calculation of indicators of destruction of the cutting tool, Bulletin of NTUU, ", KPI". Mechanical Engineering Series, Kiev, Ukraine: 66(2012). ISSN 96–102, 0372–6053 (2012).

11. Nahornyi, V.V., Zaloga, V.A.: Using vibrodiagnostics for forecasting the durability of the instrument, Izvestiya YuSGU. A series of techniques and technology, Kursk, Russia **2** (2012). ISSN 30-38, 2223-1528.
12. Nahornyi, V.V.: Method for evaluation of wear and resistance of a blade cutting tool. Ukraine Patent, UA 91817, 10 (2014).
13. Nahornyi, V.V.: Method for determining the details surface roughness in the metal working. Ukraine Patent, UA 92424, 10 (2014).
14. Nahornyi, V.V.: Device for monitoring the state of the technological metal-working system. Ukraine Patent, UA 92987, 10 (2014).
15. Panda, A., Olejárová, Š., Valíček, J., Harničárová, M.: Monitoring of the condition of turning machine bearing housing through vibrations. Int. J. Adv. Manuf. Technol. **97**(1–4), 401–411 (2018). ISSN 0268-3768.
16. Panda, A., Dobránsky, J., Jančík, M., Pandová, I., Kačalová, M.: Advantages and effectiveness of the powder metallurgy in manufacturing technologies. Metalurgija = Metallurgy. - Záhreb (Chorvátsko): Hrvatsko metalurško društvo. **57**(4), 353–356 (2018). ISSN 0543-5846.
17. Valíček, J., Harničárová, M., Hlavatý, I., Grznárik, R., Kušnerová, M., Mitaľová, Z., Panda, A.: A new approach for the determination of technological parameters for hydroabrasive cutting of materials. Materialwissenschaft und Werkstofftechnik. **47**(5–6), 462–471 (2016). ISSN 0933-5137.
18. Panda, A., Nahornyi, V., Pandová, I., Harničárová, M., Kušnerová, M., Valíček, J., Kmec, J.: Development of the method for predicting the resource of mechanical systems. Int. J. Adv. Manuf. Technol. Springer International Publishing AG, Berlin (Nemecko), **105**(1–4), 1563–1571 (2019). ISSN 0268-3768.
19. Macala, J., Pandová, I., Panda, A.: Zeolite as a prospective material for the purification of automobile exhaust gases. Mineral resources management. **33**(1), 125–138 (2017). ISSN 0860-0953.
20. Valíček, J., Harničárová, M., Kopal, I., Palková, Z., Kušnerová, M., Panda, A., Šepelák, V.: Identification of upper and lower level yield strength in materials. Materials. **10**(9), 1–20 (2017). ISSN 1996-1944.
21. Mačala, J. Pandová, I., Panda, A.: Clinoptilolite as a mineral usable for cleaning of exhaust gases. Mineral resources management. **25**(4), 23–32 (2009). ISSN 0860-0953.
22. Panda, A., Duplák, J.: Comparison of theory and practice in analytical expression of cutting tools durability for potential use at manufacturing of bearings. Applied Mechanics and Materials, Operation and Diagnostics of Machines and Production Systems Operational States – Zurich, Trans Tech Publ., **616**(2), 300-307 (2014). ISBN 978-3-03835-201-3, ISSN 1662-7482.
23. Pandová, I., Panda, A., Valíček, J., Harničárová, M., Kušnerová, M., Palková, Z.: Use of sorption of copper cations by clinoptilolite for wastewater treatment. Int. J. Environ. Res. Public Health. Basel (Švajčiarsko), MDPI **15**(7) (2018), 1–12 (2018). ISSN 1661-7827.

Printed in the United States
By Bookmasters